What Can a Body Do?

What Can a Body Do?

How We Meet the Built World

SARA HENDREN

Riverhead Books   New York   2020

RIVERHEAD BOOKS
An imprint of Penguin Random House LLC
penguinrandomhouse.com

Grateful acknowledgment is made for permission to reprint the following:
Images on page 2 courtesy of Mary Morse; page 34 courtesy of Chris Hinojosa;
page 66 courtesy of Adaptive Design Association; page 96 courtesy of LTL Architects, PLLC,
photographed by Prakash Patel; page 132 courtesy of Wendy Jacob; and
page 162 courtesy of Land Transport Authority of Singapore.

Library of Congress Cataloging-in-Publication Data
Names: Hendren, Sara, 1973– author.
Title: What can a body do? : how we meet the built world / Sara Hendren.
Description: First hardcover. | New York : Riverhead Books, 2020. |
Includes bibliographical references.
Identifiers: LCCN 2020009281 (print) | LCCN 2020009282 (ebook) |
ISBN 9780735220003 (hardcover) | ISBN 9780735220027 (epub)
Subjects: LCSH: Design—Human factors. | Barrier-free design.
Classification: LCC NK1520 .H45 2020 (print) | LCC NK1520 (ebook) | DDC 745.4—dc23
LC record available at https://lccn.loc.gov/2020009281
LC ebook record available at https://lccn.loc.gov/2020009282

Printed in the United States of America
1   3   5   7   9   10   8   6   4   2

BOOK DESIGN BY LUCIA BERNARD

In some cases, names and identifying details of subjects have been
changed to protect the privacy of individuals.

for my students—they believed me—
and for Ada, adapting

The word "building" contains the double reality. It means both "the action of the verb BUILD" and "that which is built"—both verb and noun, both the action and the result. Whereas "architecture" may strive to be permanent, a "building" is always building and rebuilding. The idea is crystalline, the fact fluid.

—Stewart Brand, *How Buildings Learn*

Disability is not a brave struggle or "courage in the face of adversity." Disability is an art—an ingenious way to live.

—dancer Neil Marcus

CONTENTS

AUTHOR'S NOTE

As this book neared press time, its central subject—where and how bodies meet the built world—became newly, vividly unresolved under conditions of COVID-19. We found ourselves venturing only cautiously outside our homes, shielded by masks and gloves, calculating every point of physical contact—in the grocery store, at the park, on the sidewalk. We also wondered about the stages of "new normal" that were yet to come—how we'd be in public space, and how we'd be together, or apart. No real certainty was on the horizon.

As I listened to the debates and timelines about the near-term future, I kept returning to the enduring invitation extended long ago by the people and stories in this book, the ones who've offered so much wisdom about the ways bodies adapt to and remake the world. Looking back and beyond—outside the latest news update, toward unfamiliar resources—can help us to look ahead, with visionary insight that transcends the present moment. People with disabilities have always confronted barriers between their bodies and the built world, and their long-held questions are relevant with a new intensity for all of us: Is a desirable future one that only restores what was lost? Or is a new set of possibilities asking to be imagined, or reimagined? The clues to how we might take up the work of adaptation—adaptation *together*—are everywhere, if we only look.

Cambridge, Massachusetts
May 2020

INTRODUCTION

WHO IS THE BUILT WORLD BUILT FOR?

A lectern for a Little Person
and a laboratory with surprises.
Where is disability? The universally
assisted body.

But one day the "why" arises, and everything begins in that
weariness tinged with amazement. —ALBERT CAMUS

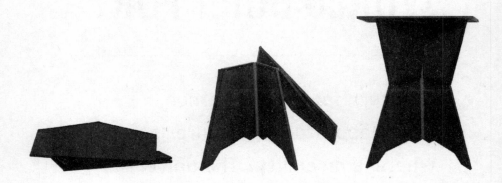

Amanda's portable lectern for short stature opens to standing position in two stages: flat-packed carbon-fiber planes unfold with hinges and are secured with hidden magnets.

Every day every body is at odds with the built environment. Bodies come up against stairs and sinks and subway platforms, sometimes with ease and grace and sometimes blundering and awkward, over hurdles, even in a sudden clash. Each flesh envelope is miraculous and mundane in this way, lugging all its gear and getting where it needs to go. Maybe you handle a sharp knife with enjoyment of its grip; maybe you wince as you sit down in or get up from your office chair. All the jostling around doorways and furniture, all the handling around sidewalks, it would be a vast and endless choreography, this daily dance of millions, if you could see us all from above. How we meet the built environment depends on both bodies *and* worlds. There's no custom-fit solution arriving for any of us, but if you could zero in on that moment of body-meets-stuff—flesh up against metal or concrete or plastics—if you could slow down the tape right at the instant of connection, you'd see it packed with information. And no one understands this more than Amanda, who came to my suburban college campus outside Boston on a frigid day in January. She was one of the first guests invited to collaborate with a cohort of two dozen engineering students who'd signed up for my class in design that semester.

Amanda is an art historian and a curator of contemporary art. She is Australian by birth and still speaks in the lilt of her native accent, even as a longtime transplant to Southern California. She has a comfortably professorial air, at ease in the front of the room in the geometric clothing that's favored by the gallery set, and she's also a Little Person. Amanda has a form of dwarfism, so her stature, at just over four feet tall, is shaped outside the standard range of average heights for humans. Her entrance to an unremarkable college classroom is a whole curriculum unto itself, because her presence has a way of casting the surrounding environment into stark relief, making some of us pause and really see it, as if for the first time: the dimensions of the space, the heights of the light switches and various media outlets, the sizes of our tables and chairs.

Amanda arrived, laptop in tow, and walked my students through a visual overview of her work as a curator, pointing to slides as she talked about all the ways she works with artists and museums to bring an exhibition to life. She told us, for example, how she'd installed one contemporary photography show at a lower wall height than standard—making it more accessible to Little People but also to wheelchair users and to children. She invited the students' many questions about how she presents her work, but also about her daily life and her experience of her body in the world, because she wasn't just a guest lecturer. Amanda was with us because she had an idea, a proposal for all of us to take up as a project— she, together with me and my eager young engineering students. She had come to ask us to design and build with her a piece of furniture, a tool that would address some of the specific features and requirements of her profession and her body: a lectern, for giving talks and for welcoming audiences to her museum shows.

A lectern, it has to be said, is so often just the architecture for a lot of hot air—that tap-tapping on the microphone that careens into feedback before settling in for the drone of voices warming to their themes and standing between you and lunch. It plays such a sturdy supporting role in so many formalized rituals—commencements, sales pitches, seminars, sermons—that, at a glance, it hardly seems worth remarking on. Think how many lecterns there are, just standing around in anticipation in the world's many hotel conference rooms and auditoriums, generic models covered in wood veneer and all more or less alike, a nondescript part of the background. Except: a lectern is also a blunt announcement, carried in the shape of an object, about who's expected to be standing behind it. A lectern assumes a world where everything is created for people whose stature ranges from just over five feet and up.

Amanda wanted a lectern at her scale. She wanted to be able to do the speaking her job entailed without a device that required her to enact the repeated awkwardness of bringing her body to the dimensions of a room at odds with her physicality, usually some structure behind an ordinary lectern. "Typically, it would be some kind of little pedestal or something, right?" she told us. It was an accommodation she'd made do with for much of her life, but she didn't want to do that anymore. She wanted a more flexible design. To present her master's thesis in graduate school, she'd commissioned a wooden model on wheels; it was gratifying for that event but it was heavy, not really portable. She'd come to our class for a new version entirely, a lectern that had to do more than stand at the correct height. She wanted to take it with her when she traveled for work, so it had to be built in such a way that it would easily fold to a flat shape, and then open back up in elegant and simple steps. It had to be lightweight for easy carrying, spacious enough to hold her notes, robust enough to support her laptop and a bottle of water and hold up under repeated use. Nothing even close was available for purchase.

My students were easily captivated by the sheer mechanics of the challenge. The entire classroom was clear-eyed, awake, eager to get going even on a winter morning. But there was still more to understand: Amanda would say she is disabled—not differently abled, not specially challenged, or any other similar variation.* Like many people in the disability

---

*In this book, I'll use *disabled* and also *people with disabilities* interchangeably; I'll also use *nondisabled* or *atypical* in place of *normal*. Many people use and claim the word *disabled* for the

community, she would use that term by choice, preferring it even to *person with dwarfism*. For her, *disabled* is not a derisive word. Amanda would say very plainly that she lives with the disabling conditions of the world. She must bring her body to the built environment, with workarounds, in dozens of ways every day, and the qualities of that interaction, body meeting world, are what render her disabled. She finds, too, that *disabled* as a descriptor connects her experience to other people with bodies that don't easily match the built world, bodies both like and utterly unlike hers. It's a subtlety that my nondisabled students had to consider at length to understand, and that my students with disabilities recognized in themselves, whether visibly or not. But the unexpected language guided our unexpected project. The task in front of us would not be a tool for assisting Amanda's body with the room. It would be the opposite—a tool for bringing the room, provisionally, to Amanda.

My students had signed up for my class expecting, quite reasonably, to use their engineering skills in a straightforward manner: designing and building prosthetics or assistive technologies for people who need them. Prosthetics or assistive technologies—tools and devices made for people whose bodies fall outside the established range of normal functioning. This would be their chance, they'd assumed, to apply all the skills they'd

----

same reasons Amanda does: not to describe the state of her body but to keep alive in the cultural consciousness the enduring reality of a disabling *world*. Others prefer "person-first" language to emphasize the human over the diagnosis. I won't be using the terms *handicapped* or *special needs*, for example, because of their locus of the "problem" of disability on the individual person, but of course they're used by others. My usages are the ones I learned from my many disabled friends and mentors, but they're not the only ones! Language is a tricky and evolving thing; it's a significant part of how cultures tell stories about people to one another. It's also just one modest site of political change and not a subject of policing I spend much time on. For nondisabled readers who are looking for a guide to their own terminology: the best you can do is to humbly and politely ask how disabled people describe themselves. And more than that: instead of fixating on getting the terms right, locate your energy and time toward seeing your own life, no matter its embodied state, as intimately tied to the strong work of disability advocacy. How might a new understanding of disability change your life?

gathered thus far—all the mathematical equations they'd learned to run, all the fabrication practice they'd had in the wood and metal shops, all the principles of mechanics that they knew as the gorgeous underlying grammar of the physical world. In this first discussion with Amanda, their minds were already racing: Had she considered something inflatable? Or with a pop-out frame, like a tent? Building stuff, getting their hands dirty for a good purpose—this is what they had signed on for.

But they couldn't have anticipated the presence of Amanda herself. She commands the room like the experienced public speaker she is, and here she was, at ease in her own body and presenting us with this singular request. She wanted a product that was useful, yes, but its requirements weren't just a technical list of needs. The request also came from Amanda's wishes. It arose from her imagination—from her sense that the shape of the world might, in a small way, be made more flexible. Contrary to the students' well-meaning assumptions, it actually wasn't a prosthesis that she wanted, at least in the strict sense of a medical device. Instead, she was presenting us with an invitation to collaborate on a material object to suit a particular situation, one shared by relatively few other people: a lectern for short stature. A bespoke design for one person, at least at first glance.

D esigners work from what's called a *brief*—a challenge presented to them by a client or collaborator with a more or less straightforward goal. It's a description of what's required at the end of the collaboration: a building, a playground, or a product, for example. You can call the designer's task a "problem" to solve if you want, and plenty of people do. But tackling design as a matter of *problems* misses much of the point. At its best, a brief is packed with *questions* that can be addressed by any number of methods. A brief isn't just a recipe-style checklist. It's a horizon, an imagined result, and an invitation for working toward that end, with a high degree of openness as to how the work gets done. That openness to interpretation can be an uneasy experience, but it's this kind of generative encounter that I actively seek to set up for my engineering students. When the work of a design team begins, across messy tables strewn with sketches and coffee cups, amid the building and the talking, there's a challenge before us, and there are lots of roads we could take to get there. We have practical and aesthetic choices to make. Which way is the best way to make an object that functions but one that also *counts*, one that can elicit a story bigger than its parts? How will the object work, but also: Why will it matter? And where shall we begin? Design professionals, drawing on several fields of expertise, say that design should address a mix of what historian John Heskett summarizes as "utility and significance." On the surface, that's an effectively succinct and even commonsense way of saying that the stuff in our everyday lives should (1) exhibit a workhorse pragmatism that's (2) simultaneously packed with expressive qualities. But consider: this dual job that design has to do is a mammoth task! How do you make a *charismatic* thing—not just a thing that works, but a thing that has elegant presence or pleasure in its handling, some kind of draw, a thing that pulls you in or makes you think while also being handy, modest, even garden-variety in its value? This combination is what makes design so interesting to so many people. It's not just the quest for a better mousetrap, and it's not just a free-form experiment, and it's not just a slick new color scheme for a mobile phone case or a toaster oven. Design calls for all of those things together: trade-offs and opportunities for mixing utility and significance. It's a mix that's hard to get right.

The brief from Amanda was relatively specific, but it was a surprising one in the context of prosthetics design. My engineering students had come to class with technology on their minds. They were drawn to things like artificial limbs and corrective gear for sensory disabilities—hearing aids, for example, or sophisticated navigational canes. From our early discussions, it was evident that they imagined themselves making impressive technical tools that would perform a curative restoration, a gift exchange for the people we work with, and an encounter from which they'd learn a little something along the way.* Naturally, in my classroom and laboratory, we celebrate those kinds of prosthetics, too. We read about and interview people who use the most technologically novel kinds of gear available—top-speed text-to-speech software that people who are blind use to read their email, or the latest in wheelchair design. It's an engineering class, after all, and technology is both our first language and our inherited course of study. But there's an extra helping of warm-hearted heroism attached to the stories about assistive technology—an emphasis on the gee-whiz quality of machinery, sentimentally wed to the experience of disability. So it's easy for my students to conceive of their task with engineering at the center, creating tools that they feel sure will solve *problems*—that will remedy broken or deficient bodies, that may very nearly save people's lives—that seem, by implication, capable of forestalling death itself. That's what makes the stories of these technologies so powerful: they fly the flag of their help—their assistance—way up in the foreground.

But Amanda's brief was irresistible to my students and me in part because it upended all those expectations. She wasn't looking to be fixed—not in the least. Her lectern would operate as assistive technology if you chose to think of it that way, but its assistance would meet a set of wishes that only she could generate. We were invited to design an object that fully captured all the energy of our imagination—something between a tool and a piece of architecture that roundly rejected the standard dimensions on which the built world is generally arrayed, and proposed instead a way for Amanda's body and the room to meet each other differently.

----

*Many disability scholars have been critical of an overreliance on what Alison Kafer calls a "curative imaginary," which she writes is "an understanding of disability that not only expects and assumes intervention but also cannot imagine or comprehend anything other than intervention." Kafer and others understand that yes, of course, cures may well be wished for, in some circumstances. But not always. Imagining disability as a richly dimensional experience has yet to enter the cultural mainstream, a topic I explore further in the "Clock" chapter.

Perhaps my students took up the project with the sense that it was a thought experiment, an engaging task for a memorable semester. But over the weeks and months that followed, they learned that the import of the brief went well beyond Amanda. The object would be one of a kind, but the project as a whole carried a hum of subtext that made us think about all of our bodies, each of us in the lab, with a new and productive strangeness. Amanda was enacting a question and teaching us to ask it also: Who is the world designed for?

The idea of normalcy—a normal, average body or mind—is so ubiquitous and mundane that it's settled into sleep in much of our collective cultural imagination. But its history as an idealized standard for human life is much more recent than you might imagine. To be "on track" in the normal curve of development—say, in the pediatrician's office, where children are measured in the hopes of landing in the exceptionally average zone—is a uniquely modern phenomenon not more than a couple of centuries old. Starting in the early nineteenth century, social scientists began the practice of collecting and studying information about populations, driven by speculation about how these measurements could be useful, especially in medicine. Where could statistics give doctors insight and understanding about people and their reported maladies, especially as mapped on a "bell curve" that helped researchers identify traits that were common—and thereby "normal"—or, conversely, uncommon, possessed by people we might now call outliers?

"Before the nineteenth century in Western culture, the concept of the 'ideal' was the regnant paradigm in relation to all bodies," writes disability scholar Lennard Davis. "So all bodies were less than ideal." In the absence of a norm, any human body was just a shadow of the admirable perfection of superhuman bodies—the ones gods and goddesses or other hero figures possessed. The emergence of modern statistics shifted the point of comparison from a lofty abstraction that no one was expected to achieve to a side-by-side analysis, assessing "normal" by observing people relative to one another.

This familiar, comparative idea of normal is so common that perhaps it feels timeless and universal, but it wasn't until around 1840 that the word was even used to describe human qualities in European languages. (Prior to that time, *normal* referred to being perpendicular or *square*, a technical term that would have been used, for example, by a carpenter.) French statistician Adolphe Quetelet adapted the practice of scientific averaging—used by astronomers to minimize errors in measurement, for example—to the science of human traits made into statistics. Quetelet claimed that his idea of *l'homme moyen*, or "the average man," could be measured and therefore ranked, both in physical and in moral qualities. Historians Peter Cryle and Elizabeth Stephens tell us that over the course of the nineteenth century, research on human populations honed its focus on this average man, such that "when the normal type was taken as a reference for the population as a whole, it served to narrow the group to be studied." There was a ripple effect to this narrowing, they write, with big consequences: "Once a rule of normality had been applied, a regular distribution could be expected to appear, producing something that had the shape of a 'natural'—and not just arithmetical—average."

Understanding the generalizable qualities among groups of people is useful in the social sciences for predictive power, and that generality has been one legacy of "normal": statistics are necessary to make sense of the seasonal tides of influenza or decide the optimal management of traffic patterns. But the habit of statistical thinking, broadly applied, creates a distancing effect, obscuring the specificities that also matter. Social scientists themselves call this the *aggregative fallacy*—erroneously assuming that what's characteristically true for a group should be inevitably true for any given individual in that group. While facts that mark a group are valuable, statistics tell us nothing about the lives of individual people.

We live and debate this tension every day: Is it more important that we see ourselves in our singularity, our uniqueness? Or that we find ourselves recognizably part of a group? Individuality and collectivity—the one and the many—both ideas make meaning in our lives, as private people and as citizens.

It took the whole of the nineteenth century for normalcy to shore up the pernicious force of a broad and encompassing mandate, a legacy that has been passed on to us: the average conjoined to the desirable, the most common range of height and weight and other qualities considered not only good but obligatory. Ideas like Charles Darwin's "natural selection" were culturally (and mistakenly) interpreted as hereditary directives from nature—a normalcy that could be identified, and then perhaps overtly enhanced and encouraged, with *average* taking on the conceptual heft of *better* and *best*. Quetelet's legacy was to make the average, what started as unremarkable by definition, into a paradoxical ideal. When the average is laden with cultural worth, everything changes: what was common began to be seen as what was "natural," and what was "natural" came to be seen as *right*.

The cult of normalcy reached its ugliest expression in the eugenics movements of the early twentieth century, with an aggressive pursuit of normalcy that brought violent measures for disabled people: mass sterilization campaigns and euthanasia, to ensure that only the "right" kinds of individuals and family units would genetically flourish, in the hopes that nations might flourish as well. This wasn't a sinister idea promoted by a few thinkers on the fringe; eugenic thinking spread in the 1920s in the United States in utterly benign-seeming forms, like genetic "contests" held at state fairs in the Midwest. In them, "Better Babies" and "Fitter Families" were evaluated for their relative heritable qualities, not unlike the competitions among livestock and garden specimens that were staged nearby. Signs that promoted these contests made the high stakes for an optimized population unmistakably clear: "Some People Are Born to Be a Burden on The Rest." In nations eager to promote the overall health and thriving of the "we," eugenic thinking superimposed the value of correctness and acceptability—even civic responsibility—on the concept of normalcy.

Even if the darkest chapters of this history are behind us, the tacit equation of normalcy with human progress and perfectibility has settled quite comfortably into our daily speech and rote decisions. The aggregative fallacy shows up every time we assess whether we're "measuring up," whether we have what it takes to "get ahead," whether we're demonstrably on or off the charts, by comparison with others. Davis sums up the impact: "The introduction of the concept of normality . . . created an imperative to be normal, as the eugenics movement proved by enshrining the bell curve (also known as the 'normal curve') as the umbrella under whose demanding peak we should all stand. With the introduction of the bell curve came the notion of 'abnormal' bodies. And the rest is history."

The rest is history! The most repugnant consequences of the eugenics movement may be in the past, but there's a dotted-and-dashed line from its overt rankings of humans to the everyday contemporary habit of organizing around desirably normal human traits and behaviors. Witness the milestones with which parents anxiously chart the development of their infants, and the elaborate ranking and sorting of young people via the scores produced by high-stakes testing. The pursuit of exceptional normalcy belies the quick and unquestioned work of two distinct ideas, now conflated: that being robustly average—just "ahead of the curve"—also means having the best shot at a good life.*

How strange, then, to land squarely in the present and reckon with what is also true—that according to a 2011 comprehensive report by the

----

*Normalcy as a word and an idea has a much more complex history and set of meanings than I'm able to do justice to here. See my citations for further reading, but also see Rosemarie Garland-Thomson's notion of the normate—how normal becomes the default cultural experience of the mythical independent, able-bodied, ruggedly individualistic self, in her book Extraordinary Bodies: Figuring Disability in American Culture and Literature.

World Health Organization (WHO), one billion people, an estimated 15 percent of the global population, live with disability—people whose operation of body or mind falls well outside that canopy of normalcy. A full billion people who, like Amanda, have bodies that defy a world designed according to a rulebook of standardization. The report includes motor and sensory disabilities, mental health conditions, cognitive and developmental disabilities, and general aging. These conditions may be hereditary or acquired, a result of poverty or happenstance, dependent everywhere on context, experienced differently when organized by race and gender, but united in the report by a powerful idea: normalcy is not quite so overwhelmingly dominant as it might appear. The numbers in the report suggest that disability is a common part of human life—an ordinary experience, infinite in variety, replete with creativity and heartbreak, from sources internal and external, and carrying social stakes everywhere.

Although the report takes stock of our collective state of disability, it is not primarily aimed at labeling and counting its innumerable instances. Rather, it points beyond the body to include the "disabling barriers" that stand between bodies and the built world, especially bodies closer to the tail ends of the bell curve. Considering what's made possible or impossible by technology or architecture shifts the focus from "abnormal" bodies to a "lack of accessibility"—or, in the case of Amanda, from *short stature* to *lecterns everywhere*. It's the absence of things like ramps and hearing aids and special education software that makes the barriers to human thriving higher. The report expresses a long-held insight articulated in the scholarly field of disability studies: that ability and disability may be in part about the physical state of the body, but they are also *produced* by the relative flexibility or rigidity of the built world, its capacity to bend or adapt in a dance with bodies in a range of states and stages. Disability in part *results* when the shape of the world—buildings and streets but also institutions, cultural organizations, centers of power—

operates rigidly, with a brittle and scripted sense of what a body does or does not do, how it moves and organizes its world.

Disability studies identifies two mental models that serve as useful contrasts for understanding these relationships between the body and the world. In a purely medical model, the body is the location of impairment, which suggests that the person with the impaired body bears the responsibility for it—for telling a story of coping, or surviving, or overcoming, or any number of other possibilities, all of which require the individual person to contend with a *personal* condition. A social model of disability, by contrast, invites you to widen the scenario from the body itself to include the stuff around it: the tools and furniture and classrooms and sidewalks that make it possible or impossible for the body, however configured, to do its tasks, and the larger structures of institutions and economies that make human flourishing possible. In a social model, it's the *interaction* between the conditions of the body and the shapes of the world that makes disability into a lived experience, and therefore a matter not only for individuals but also for societies.*

The condition of disability is present whenever a body finds itself in what scholar Rosemarie Garland-Thomson has called a pointed "misfit" relationship with the world—not the melodrama of a tragedy to overcome, not merely a "defect" of the flesh, but a misfit: a disharmony that runs both ways, body to world and back. Garland-Thomson is a pioneer in disability studies professionally, but she also understands misfit status

----

*For more on the basics of the distinctions of these models, see, for example, Tom Shakespeare, *The Disability Reader: Social Science Perspectives*. The medical and social models are not simplistic opposites and have received much more nuanced attention and theory in disability studies, but the basic distinction remains useful for anyone thinking about bodies meeting the world. In citations throughout this book, I have tried to point those who are interested in and new to ideas in disability studies toward ideas for further reading. In the text itself, in the interest of economy and accessibility, I've tried to indicate complexity while also maintaining brevity.

in her own body. She has two atypically shaped arms, hands, and sets of fingers, so she lives between a body and a designed world that make her job and her life both frictionful and, by necessity, deeply and flexibly adaptive. She lives life—eating and drinking and opening doors and using the voice-activated features of her smartphone—as "a square peg in a round hole," which is different from understanding her body as *broken*. Simple medical diagnostics could never fully capture her particular tango with dishes and doorbells, to say nothing of composing her short emails and long books using voice-dictation software. Misfitting encompasses not just the shape of her body, but the shape of the world, too, together and also working against one another—peg and hole mutually at odds, as the saying suggests, and then perhaps arriving at other, creative ways to work in something like concert. But *misfitting* nonetheless. It's a deceptively casual word, even something like slang. Misfitting provides a bracing shorthand to stand for all kinds of bodies in a clash with the built environment—the obvious collisions and the quieter ones, hiding in plain sight.

Here is a partial misfit inventory from my own tiny sphere of everyday life: My upstairs neighbors, two widowed sisters, each insist that the other is in need of hearing aids. An injured colleague finds a first clumsy, then clever "new normal" way of fumbling around the kitchen and the office with a wrist sling. The child of one acquaintance is diagnosed with scoliosis that necessitates a back brace; the child of another is diagnosed with acute anxiety; a third awaits a designation along the autism spectrum, emergent and tricky to pin down. A friend from high school monitors the slow creep of her father's glaucoma from a thousand miles away. A teenager's persistent pain turns out to be rheumatoid arthritis. A septuagenarian architect lost a finger in a woodshop accident, a first grader's stuttering became pronounced, and three siblings got locked in a year-long legal battle over their mother's brain death. That family had to grapple with the very meaning of aliveness—capacity, ability, sentience itself—before deciding to unplug the machines, the assistive technologies that kept her breathing.

Who is the world built for? That's the question in the room when my students and I are creating the lectern with Amanda, a deceptively modest and singular brief that carried with it the freight of a global-scale investigation. Amanda could easily have continued to use boxes or steps or some other assistive aid to reach the scale of a standard lectern, but she was interested in something more than a product that would get the practical job done. Utility, in other words, was crucial, but the *significance* of the design was also at play. Her idea for the object came from a zoomed-in, microcosmic look at the moment when a body meets the built environment in a misfit. But Amanda wasn't looking for a form of rescue from the state of her body. She wanted this lectern to do its work not only as a solution but as a question, a *designed* question in material form: Who fits in and moves through space? Who gets into the room, moves through the door or down the street, yes, but also—who gets the education, acquires and keeps the job? Who steps up to reach the microphone, the one saying *Listen to me?*

In some ways, my own presence in that classroom was more surprising than Amanda's. I'm trained as an artist and writer, and in school I'd been the kind of student hastily shunted into classes like Physics for Poets instead of tracked for science, technology, engineering, and mathematics—the bundle of national priorities known collectively as STEM. I made it well into adulthood thinking that engineering was only ever about building things in a straightforwardly practical way, full stop. I thought "technology" was just the *how* of the way things worked: the means by which science and math, via their respectable but rule-bound outputs, determined how strong, how fast, how efficiently a thing could run. To be frank, I thought engineering was a brute-force and impoverished way of seeing the world. I thought technology had little to offer the big, first-principles questions about life on Earth that preoccupied me—the *whys* of human existence, the ambiguity and poetry I thought only writers or artists, equipped with their powers of imagination and expressive tools, could take on.

I was wrong, of course, about engineering. Nothing is more imaginative than the laboratory when ideas are under construction, when the possibilities and questions are truly open. At its best, there's a hunt for discovery, a rhyming energy to the artist's studio. I stepped into engineering by necessity—because there *is* a brute-force and impoverished definition of disability at work in the world, a perennially underimagined experience of human life. Better technologies are necessary, but they're not sufficient. The condition of disability is too various, too interesting, and too urgent to greet and address with any single domain of research. To reconsider all those long-calcified ideas about normalcy, my lab—guided by people like Amanda—needed the workhorse guarantees of engineering, but also the provocations held up in art and design. My students and I have helped design an enormous ramp—not for accessible entry to a building but for a stage, for a choreographer who wanted to use its physics for dancing in her wheelchair. We've worked with a special education class to make soft felt furniture for kids, with a one-armed man on a prosthesis for rock-climbing, with a blind artist who asked us to turn his cane into a musical instrument, and much more. Disability is too interesting to have to decide between pure practicality and raw beauty. Misfit states beg for art and engineering and *design*. It was design that allowed me to insist on a world that included ordinary stuff that works, but also stuff that *sings*. Utility and significance, solutions and questions, held open. Nothing about engineering is my natural language, but I have learned to hold its vocabulary in my head and in my mouth, making my peace with an outsider-insider place, earning a slow and hard-won expertise.

Getting to that classroom wasn't just an intellectual exercise. Some things we understand in our heads, and some we come to know in our bones. I come from an atypical extended family—though perhaps every family is just that? A couple of us are on the autism spectrum, and we've got dyslexia and chronic depression represented among us. Misfit states were common enough in my experience to seem natural, and even, at times, invisible. But the real catalyst for my education arrived with the first of my three children, my son Graham, who was born with Down syndrome. To remember his birth is easy, an adrenaline-spiked charge of memory, but remembering the foreignness of the diagnosis all those years ago is murky now. I have to consciously wind the tape back to unlearn what would become its ordinariness in our lives, to remember there was a *before*. Graham's birth came with a whiplash of highs and lows for my husband, Brian, and me: the peak miracle of a newborn's arrival and the deep valley of grief—though even now I resist writing the word *grief*—brought on by the inescapable reality of the misfit. It was at times impossible to contain the thunderstruck new normal of those emotional ups and downs: a child whose wholeness and joy lit up our days and simultaneously brought well-meaning, ham-fisted condolences from friends and acquaintances. Even in Graham's infancy, before we had the words, there was a question in our tiny house of three: Who is the world built for? We were already in love with our child. He was wanted in every way. And we were immediately conscripted into an unsought, sobering education about the way bodies meet the world in unavoidably misfit collisions. Assistive technologies— the fact and the invitation to a family structure newly, acutely, permanently marked by *assistance* itself—would change the rest of our lives.

But it was early days then. Long before that January morning in class with Amanda, long before I could even imagine a life in engineering, I walked into a homogenous box of commercial space in a Pasadena, California, shopping center, this one converted into a gymnasium called the Center for Developing Kids. Graham was on my hip, not yet a year old, and engineering education was far from my mind. We were there for routine physical therapy, one of dozens of appointments we had in the early years.

This one involved what I suppose you'd call gross-motor "exercise," at least of the kind suitable for tiny humans, though Graham only ever experienced it as play. There were swings for the sensation of moving through space, and balance beams low to the ground, and colorful soft foam mats for rolling around with all manner of cleverly disguised gear, toys and tools strewn everywhere as props. These objects were meant to challenge a small person's body to do things that didn't otherwise come naturally: bend or stretch, if needed, or strengthen and align, as in my son's case.

In Down syndrome, low muscle tone is global in the body and nearly universal. So much "early intervention" is designed to counteract the delays that tend to come with that weakness—late crawling, late walking, late running—and the overall cognitive delays that can be exacerbated by a lack of movement. All play for all babies is developmental, of course, but at this gym, the toys were calibrated to be irresistibly colorful and joyous but also precisely, incrementally challenging. If Graham wanted to grab a block some distance away, how could we set up the structure so that it was within his reach but also testing his wobbly balance? This was approachable design for bodies whose capacities fell far outside the developmental charts.

I was living with so much worry about raising a child with big differences—deep in a ricochet of conflicting impulses about how best to love him, and most days unable to ask for the help I needed. Should I research obsessively, or focus on the day-to-day? Should I tell my family about the hardest parts, or reassure them with pictures and videos of him in tiny onesies? Right in the middle of all the late-night online searches and diaper changes and doctor visits, there was also this weekly trip to the gym, and inside it was a riot of gadgetry. Here was more imaginative prosthetic hardware than I could ever have thought possible: bouncy balls, sensory chew toys, stretchy pressure vests and weighted blankets, the tiniest possible ankle braces, and so much more that would be dutifully recommended to us by the experts. A zillion extensions and appendages that posed the same question: What can a body do? What can it do now that was impossible last week? What can a body do with and without the designed world around it, now, and in an hour, and later this month, and in two, five, ten years? Every day, Graham was becoming more singularly himself, showing his interests and sense of humor to Brian and me—becoming an individual, not a diagnostic type. But the world was not designed for him, we knew. That is, we knew in spurts and then suddenly all at once, because we saw this square-peg, round-hole conundrum. We were clear-eyed about the misfit condition of our interconnected lives for the long haul, but we were unclear about the *next* right question. Should we encourage him—his body, his skills—toward patterns more aligned with the world? Or could we ask the world, in part, to flex and bend some of its structures around him? Misfitting carries a charge to both individuals and collectives alike.

In the waiting rooms and hallways of all those appointments at the gym, with time, I started to see the gear for other kids, too. These were tools for all kinds of bodies, not just for people with Down syndrome. As an artist, I was trained to look closely at visual culture, to attend to all the stuff that's made. And here were these tools, the visual embodiment of everything that Graham, a baby with the tiniest possible glasses secured by a short length of brightly colored elastic around the back of his head, was teaching us then—that bodies come both beset and enhanced by infinite complexity, and that a whole world of tools can variously bridge the awkwardness between bodies and the landscape, or, alternately, the hardscape. It was a newly visible misfit choreography, a meeting of body and world that changed everything.

Graham also brought to my life so many relationships with *people* I would never have met otherwise. I could not have imagined then, as a new mother, how over the next decade my social life would fill up with people who are blind or deaf or use wheelchairs—and whose everyday routines were also replete with tools. Tools for assistance, more interesting and various than I'd ever encountered. All the canes and walkers, the portable oxygen tanks, all the crutches and heavy-duty adaptive tricycles, the extra-support strollers, the plastic syringes for feeding. All the detritus for bodies mismatched, the assistive technologies that do the spotting and fortifying for all the soft flesh of the world. I might have missed it—if not for the role of parent.

Graham needed very practical gear, but our family in its misfit states would also need much more than good solutions for the short horizon of the here and now. An uncertain future lay in front of us, and the objects at the gym seemed to tell that story—what was to come—and, simultaneously, to suggest all the ways that our prospects were unfixed. Our family's future would be *built*. It would expand or contract depending on human capacities, yes, but also depending on the built world. Once my attention was trained on the vivid work being done by these tools, I could no longer unsee how alive and redolent with meaning they are. The right tools and the wrong tools at the right or wrong time: blundering and leaden or versatile and elegant or somewhere in between. I found I had to be inside that story—part of how those tools are made, part of a laboratory for the work of design.

I was already an experienced teacher by the time Graham arrived. I knew the way a classroom gathers a group of strangers into a provisional *we*—the deceptively ordinary, actually magical encounter made of an agreement to meet at a regular time and place and bat around ideas together. The energy of a classroom can be like a cerebral think tank or like a lightly buzzing party about to go off the rails. It's different every day. The dyads that get created, two-way relationships between teacher and student or between students, can make or break the experience. But the *triad* that gets made in a design class—you and me and this concrete thing we're building together—I couldn't have imagined it when Graham arrived. A joint project, the something-from-nothing in an object you can knock around, has a collectivizing effect by its repeated acts of giving form and shape to an idea, a thing under construction. For me, as in the project with Amanda, the pleasure of it surpasses even the best discussion made solely of words.

After her visit, my students' questions for Amanda were multiple and immediate. How did this unusual request fit alongside the engineering they were accustomed to—this request for a one-of-a-kind object, not destined for mass manufacturing? How should they talk to Amanda about her bodily stature, much less her wishes for this strikingly personal piece of furniture? She had introduced them to some basics, pointing them to the advocacy work of an organization called Little People of America, and teaching them that "leggy" is a shorthand term among Little People for folks of average stature. But there remained considerable awkwardness to navigate. In an engineering laboratory like mine, we design and build things as a social practice, and we start with questions that are never just technical. Engineering is not the science of the laboratory alone, after all. It is fundamentally applied, which means its results live in the world. It belongs to people, not just as "users" but as the protagonists of their dimensional lives. So, students wondered, could they ask Amanda what was also on their minds—about her daily experience of the world? Questions like: How did she shop at a grocery store with high shelves? How did she drive a car? And was it okay even to say the word *dwarfism* in the twenty-first century? They wondered if they could ask about all those images they'd seen in the media of Little People, in the ugly histories of so-called freak shows and the shrill contemporary light of reality television. Would it be hopelessly rude to bring up these subjects? Over weeks and months, the goodwill and conviviality of designing things together made it possible to broach questions like these. A shared material task will direct everyone's gaze to the work at hand, but good collaborations are structured by relationships that you can't just assume. They don't arrive without effort. They're also built.

I assigned four students to Amanda's team, and they set to work researching their options for the design. They pushed together some tables as a workspace and pinned dozens of images on a board nearby for inspiration. They made a bunch of tiny scale models out of cardboard and paper, all manner of small geometric shapes, three-dimensional sketches of various kinds of hinges. They were trying to envision a folding pattern for this lectern, the series of steps by which it would open and close in a way that would be easy to use, and not hindered by cumbersome extra weight. It had to be soundly engineered to take a lot of wear and tear, but it also had to have a distinctive appearance that would match Amanda's imagination for it. They kept in mind that she was interested in an object that would not only perform well in a nuts-and-bolts way but also upend her audience's expectations—about the room as a platform in a dual sense, and her role in it.

The students considered their options and started prototyping, trying ideas in material attempts that they'd test, reject, edit or alter, and test again, rather than just discussing the possibilities and debating choices in the abstract. They showed Amanda their models over video chat, explaining their operation and assessing their merits, mining her feedback for ideas they could take back to the lab. The images on their pinup boards doubled, enclosing their worktables in a makeshift house, with walls that mapped out raw potential: pencil sketches and computer-assisted design renderings, magazine clippings, and still more tiny scale models. They were getting nearer to the crux of decision that comes with every engineering or design process. It never gets easier, as designers will attest. They had to figure out how many weeks they could keep the possibilities open, and it's rarely as long as a team would hope. They had to commit to an idea and bring it into being.

By the time spring arrived, my students had delivered to Amanda the model they'd agreed on together, a sleek angular lectern that unfolded in three moves—first the hinged planar legs, then the side supports, and finally the "desk" portion on top, swinging around and resting as the lid of the geometric base. It was made of super-strong black carbon fiber, a material most commonly used in aerospace and automotive engineering, with a woven structure that, when cured with an epoxy solution, has a fantastically high strength-to-weight ratio. The lectern held internal magnets to secure the places where the folds met, and the students added sturdy rubber edging to protect it from travel damage. It folded neatly into a modest stack of flat planes when not in use and, when opened back up, was sure to interrupt the humdrum operations of every space, from spare modern galleries to the dreariest of partitioned conference rooms. This out-of-sync object would throw all the standardized dimensions of the built world newly into relief, raising questions just by its arrival. Amanda was thrilled, and the students were proud. The project had powerfully stretched their fabrication chops as they took it from simple two-dimensional sketches to a functioning product in a semester. And it had done something more for all of us: it had stretched our ideas about the larger project we'd enacted in a class that was nominally about "assistive technologies." In the course of Amanda's project and others like it, my students started seeing forms of assistance everywhere, including in their own lives.

What is any technology doing, any tool, any implement, if not offering assistance?* It is the very nature of all of our stuff to give us help. Think now of all the ordinary objects that extend a body at any point in the course of a day: eyeglasses, knife and fork or chopsticks, perhaps a walking stick for hikes or a plastic arm that can throw a ball for a dog. Consider the infinite extensions and outsourcing of tasks that happen

----

*Katherine Ott and other scholars have pointed out that this phrase is redundant: "Since all useful technology is assistive, it is peculiar that we stipulate that some devices are assistive while others need no qualification. Besides serving to stigmatize and segregate a benign and inanimate entity—a device or appliance—the term 'assistive technology' also needlessly complicates understanding of the devices so designated" (Ott, 21). Still—"assistive technology" is the shorthand most commonly in use, and it's unlikely to go away. I often use the term *adaptive technology* in this book and when training students, because its meaning is more precisely true.

via mobile phone, whether augmented by "smarts" or not. Open your kitchen catch-all junk drawer: paper clips, toothpicks, elastic bands, pushpins. These are the commonplace prosthetics and assistive technologies that are at home in the world with all our many bodies. Tools for holding the world intact when it threatens to fall apart in a mess—tools for reaching, bracing, connecting, the low-tech and the high-tech kind, together. They're all no more and no less than assistive technologies, the things that a body uses to make its way in the world. Amanda's lectern, seen in this abundant context, is a tool that assumes its place in the giant family of designed objects that are crucial for everyone.

Tools are not just the body's extras, hanging around at the ready for moments of special assistance. They are the ordinary and necessary amplification of the body itself. The body is made of organic shapes— rounds and swells of joints, muscles arranged as weights and counterweights strapped together with ligament—and all of it covered with the soft, porous mega-organ we call skin. It's a miracle in every form, just as it is, and it also needs stuff to carry out its plans. The tools we use, and the environments in which we move, are built to compensate for our bodily limitations or to refine our capacities: how hard we can strike a nail or slice through a vegetable. There's a reason historians locate some key origins of human civilizations in the periods where deliberate tool use is evident. The stuff we use between our bodies and the natural or built world—these augmentations are how humans organize and get life done.

A body is almost never not-extended. You might even say—and there are plenty of philosophers who do, often at length, sometimes at lecterns—that the human animal is co-extensive with its tools. It's the most remarkable and unremarkable thing in the world: a body *plus* its many things, *plus* any number of augmentations, from the most elemental to the most mechanically complex. Whether you are among the billion who, living with distinctive disabling conditions, extend their bodies with prosthetic gear, or you're merely among the daily wearers of contact lenses or orthotic shoes, your everyday life offers nonstop evidence that the body-plus may actually be the human's truest state. Look again at bodies confronting the shapes of everyday life and you'll see, as Davis observes, that "the seeming state of exception of disability turns out to be the unexceptional state of existence."

Davis also writes this admonition, however: "It's too easy to say, 'we're all disabled'"—too easy and a gross generalization to mask the specific realities of misfitting that arrive for some bodies and not for others. My students and I, each of us at average stature, had not experienced the world as Amanda did. Rather, doing design with Amanda allowed us to see, via her disability—her misfit condition—more deeply into her expertise, into the nexus where the body meets the built environment. It allowed us to join her in a shared task of building and rebuilding a small part of that world along more deliberate and creative lines. Amanda helped us see that the opposition she experienced at the juncture with a room—her being "too short"—was in part a diminishment in her experience, a *closure* that necessitated assistance just to allow her to move through the world. But her ingenious, adaptive perception and her design brief simultaneously made possible a stunning and urgent *opening*, too.* In the pursuit of utility

----

*This idea of *openings* in disabled experience has been beautifully articulated in disability studies. "I do want to claim that, collectively, we have accumulated a significant body of knowledge, with a different standpoint (or standpoints) from those without disabilities, and that that knowledge, which has been ignored or repressed in nondisabled culture, should be further developed and articulated," writes Susan Wendell in *The Rejected Body: Feminist Philosophical Reflections on Disability* (73). "Moreover, people with disabilities have both knowledge and ways of knowing that are not available to the non-disabled. Although I hope that their knowledge will ultimately be integrated into all culture, I suspect that any culture that stigmatizes and fears disability would rather ignore and suppress that knowledge than make the changes necessary to absorb it. It may have to be cultivated separately until non-disabled society is transformed enough to receive and integrate it" (Wendell, 75). See also Rosemarie Garland-Thomson (2011) on this idea: "Acquiring or being born with the traits we call disabilities fosters an adaptability and resourcefulness that often is underdeveloped in those whose bodies fit smoothly into the prevailing, sustaining environment. This epistemic status

and significance, design can include making something new, but it can also include *unmaking* the world as it is, or perhaps *remaking* it, with parts and systems alike. When Amanda uses the lectern now, the newly altered scale of her furniture makes the room she's in contract a little—it adjusts to her size in your mind's eye, if just for a moment. When she's done speaking and collapses it to its portable stack, the room resumes its normative average shape. That's why the lectern does such effective work, as a question in the form of a product. What *if* the features of this room were to become another size than they are, a new version of themselves? If we could work with Amanda to make a new kind of lectern—well, what else was out there, awaiting reconsideration and redesign? My students learned to see their own stakes in the question of this lectern, to consider the ways they or their loved ones variously misfit, a condition more on a continuum than they'd previously thought. Closures and openings lie all around us—in implements and furniture, the architecture of a room, the layout of a street corner—where the meeting of body and world reveals a disabling barrier that might be made visible, and might then be mitigated or even, with time and commitment, dissolved. They allow us to pose the larger question that is the most resonant work of design: *Who is the world built for?* Or, to put it another way: *What can a body do?*

T his new book about design and disability takes its title from that very old question. *What can a body do?* is a famous one in the history of philosophy, with its original appearance in the work of Baruch Spinoza in the seventeenth century. The French philosopher Gilles Deleuze, in his 1992 essay of the same name, took up the question to explore the deep plasticity of the body as a set of capacities. "A body's structure is the

---

fosters a resourcefulness that can extend to the nondisabled and not yet disabled as they relate to and live with people with disabilities" (604).

composition of its relation," he wrote—it is "endowed with a kind of elasticity":

> What's more, its composition, as also its decomposition, passes through so many stages that one may almost say that a mode changes its body in leaving behind childhood, or on entering old age. Growth, aging, illness: we can hardly recognize the same individual. And is it really indeed the same individual?

Our bodies are not just the sacks of flesh that hold our "real" intellectual selves; they are not fixed entities but mind-bogglingly adaptive, responsive instruments. The moves we make are contingent everywhere on the arrangement of the body-*plus*, and nowhere is this radiant dynamism more present than in design. In an exhibition she curated with this same title, Amanda herself posited a sharper question still, one that was the engine for our project together: not "What *is* disability?" but "What does disability *do*?" That's the question in the choreography of the misfit: Rosemarie dictating into the keyboard on her smartphone and Amanda at the lectern and Graham on the bouncy ball and each of us, in turn, in the *doing*, operating between the changes in our bodies and built environments that we'll meet in a mismatch over the course of our lives.

What can a body do? was the most powerful, double-edged question delivered by Amanda's lectern. Her performance of public speaking was determined by the inherited standardized measurements of the world but also by a bridge of possibility that could be newly built. This one object, in other words, brought both the bite of critique and the possibility for repair: a frank and unvarnished exposure of the discord between a body and the world that was her daily reality, and the insistence, despite everything, that her body might greet the world in a structure made new. Neither critique nor repair alone would do justice to the dynamic state of the human body and all its many tools.

Holding the question of this book open is the only way, as a maker and a researcher and also as a parent, that I can account for the twin truths of the designed world: that material objects are an index of ideas that built its many structures, their steel and concrete often deep and intractable—*and* the provocation that some of them, at least, might be remade otherwise. Disability reveals just how unfinished the world really is, in its mundane forms and in its most aspirational politics—a contemporary reality tested most acutely under conditions of global pandemic, requiring fundamental shifts between our bodies and the world, and mutual trust despite deep uncertainty. That unfinishedness is the engine of the stories I report in this book—disabled people with their design experiences at the creative heart of adaptation, people in the active work of building and rebuilding their worlds, with insight and high stakes implicated for everyone.

How does a man with one arm tackle the job of changing his newborn baby's diaper? How do prosthetic limbs in Ahmedabad and cardboard chairs in Manhattan get designed and built alongside the people who use them? How does a deaf student use the architecture of a room for signed communication? And how does a man with a new diagnosis of amyotrophic lateral sclerosis (ALS) design a home for a desirable future, one that anticipates the features of the place where he'll want to live when his physical mobility has been reduced to almost total stillness? These and many other stories of bodies meeting worlds in this book invite you to consider your own extended body with the recognition of that same assemblage happening for all of us: adaptive and mysterious, its meaning made of how we sense and move and think, but also how we each meet the built world, now and in the future. The stories here are perhaps the most important of all—how the inheritance of normal shows up in the material lives of people everywhere, and what happens next. Which choices are the ones that matter, and would we make the same choices if and when the misfit story becomes ours? The quality of our attention to the built world determines not only who and what we see in it, but the shape of the narratives we tell about our lives.

Disability scholar Simi Linton writes that "the cultural narrative [of disability] incorporates a fair share of adversity and struggle, but it is also, significantly, an account of a world negotiated *from the vantage point of the atypical*." Perhaps that statement seems obvious at first hearing, but many nondisabled people have never really considered that disability as an experience might indeed *be* such a generative and restless vantage—a view on the world, a perch for surveying, a lens for encountering the status quo from outside its normative, taken-for-granted landscape, and an active, lively form of negotiation. Even fewer nondisabled people have considered that the vantage of disability experience is also on offer to everyone as a form of shared *knowledge*, a wealth of understanding that has formed and re-formed the very shapes of the contemporary world.

Design is one way to track this vantage and the new stories that come with it—in things we wear and use in our houses, and in the buildings and roads that are the settings of our lives. For disabled people like my son Graham, the world must necessarily be mutable—it must be seen as such or made to be so, its structures pried open and reconfigured. The stories of people and design that unfold in this book make that mutability both vibrant and urgent in what I've structured as a kind of travel narrative, starting with the limbs of the body itself and moving outward incrementally: to furniture, to rooms and buildings, to the public realm of streets, and finally to the clock—to the built expectations of time that are out of sync for people like my son, and the clocks in all of our lives that act as an index of our worth.

Amanda arrived in my classroom and called herself disabled. Perhaps you would or would not describe yourself as such. But disability is not a fixed or permanent label that belongs only to some people; it arrives for each of us. Short-term injury and long-term illness, changes in our perception and mobility (and the perceptions of others *about* us), the chronic

31

misfires that happen in our emotional makeup—if it's not a reality in your life now, it's sure to be so in some form, in your own body or among those who share your intimate life. Disability gathers a dimensional *we* like nothing else, because disability is no more and no less than human needfulness, both personal and political. That's why the *we* that ties together this book is as tenuous as it is important: the collective that arises in the form of shared bodily vulnerability, the ways our physicality and our thriving are tied.

So often, of course, the first person plural is a falsehood: Who is this *we*, a word mostly used to blithely generalize from one person's limited experience in a myopic way? Our distinctions and specificities are important. But for disability and design in this book, the *we* is both real and profound. It's not that all our bodies are the same. It's that the stakes for life together are universally shared by the misfit states that come for every body. We find ourselves in need of assistance—some of it from the forms of the designed world (or, as these stories show, the *redesigned* world), and some from one another, body to world and back. But getting help? That's for my son, Graham, but it's also for me and for you, for all of us. Not everyone should call themselves disabled, but everyone should recognize that both giving *and* receiving assistance are actions we will each take up in turn, every one of us. Human needfulness really *is* universal. We—and I do mean *we*—might choose to let tools for assistance be visible and unifying.

Looking closely at household objects and furniture, rooms and buildings, all our designed forms of help, is not a specialist's concern. It's an invitation to see both our bodies and the built world as movable parts and systems—more malleable and under construction than perhaps we realize. Who *is* the world designed for? Who, in turn, can grant or summon the power to do the designing or the building? And what can a body do—whether in life in our modest everyday, or when conditions suddenly change? Let these be our restless and generative questions, each of us in our many bodies making and remaking this adaptive life.

LIMB.

Cyborg arms vs. zip ties: Finding the body's infinite adaptability and replacing the things that matter.

Neither the naked hand nor the understanding left to itself can effect much. It is by instruments and helps that the work is done, which are as much wanted for the understanding as for the hand. —FRANCIS BACON, *NOVUM ORGANUM*

Chris uses his shoulder to suspend the ankles of his infant high above the changing table, working with a holster he fashioned from soft cords and felt. •

M ike is forever talking about his prosthetic arm, whether he wants to or not. At any dinner gathering with acquaintances, the subject will eventually come up, because people find his body, extended with its machine parts, to be an irresistible enigma. Strangers glimpse it and thank him humbly for his nonexistent military service. He's an unintentional sensation at airport security. Mike is tall and an unhurried talker, affable, like a high school athletic coach in the off-season, with an open face and thinning short blond hair. He has a flesh arm, the dominant one, on his right and wears a prosthetic arm as its match on the left: a forearm, wrist, and hand sheathed in a semitransparent hard plastic housing, pearly gray and smooth and contoured in the shape of a recognizable arm and hand, with each finger articulated as though in a full-length glove. There are seams where wrist meets hand, and tiny screws that bind the parts together, and beneath the outer "skin" you can see the inner circuitry. When the wrist or digits move, the motors make a faint whirring sound, and the entire structure has a mechanical sentience about it that strikes a dramatic departure from the conservative blue button-down shirts that Mike wears most days: the human body and its mixed assemblage.

Mike had visited my classroom as a guest lecturer on his experience with prosthetics, but I wanted to see his tools and his life outside the context of engineering. I wanted to sidestep the impulse to see Mike's limb as it's portrayed in popular culture—as a fantastical curiosity, the stuff of so many breathless cinematic narratives. So I arranged to visit him at his home in a suburb outside Boston. I wanted to talk to him about the small archive of four arms that he's variously used for decades as the material culture of his life. "Where I am right now is so different from when I was fourteen, and at twenty-two, and at thirty-six," he told me. "And it'll be different five years from now." To see and hear the whole story, it made sense to be at his house, outside the high-tech labs where these limbs are designed. We sat at his kitchen counter one day in late spring amid the sounds of construction—his driveway was getting repaved—drinking seltzer and talking about prosthetic limbs as he answered contractor questions and fielded the occasional phone call from his kids.

People say to Mike all the time: "That's the coolest thing I've ever seen." His prosthetic arm is a popular subject of conversation because his body looks so much like the future, or what you might imagine it to be. People want to know: Can he pick up just *anything* with that arm and hand? Does it ever hurt? How does he put it on and take it off? Encountering him, they exhibit an overwhelming impulse to see him as nearly superhuman in his hybridized state, to speak of him in the imprecise slang of a "bionic man," a creature that emerged from science fiction into real life. Its apparent qualities of simulacrum and enhancement are what give Mike's prosthetic its power: a designed object that looks so similar to the body in form, and so much like expensive machinery in function. The lifelike hand, with all its digits faithfully replicated, looks comprehensive and replete and therefore persuasive as an object, as though it might be strapped on at a moment's notice, taking the body into the future in a smooth exchange of this-for-that.

An amiable disposition, a comfortable professional life in sales, and longtime fatherhood have made Mike an easy conversationalist. When the questions arise, he is inclined to humor his dinner companions or the airport security staff. "I think as the technology has advanced, public perception has improved, and public curiosity has improved," he said. So, now, rather than avert their eyes, the curious walk right up. "Does that give people permission to ask?" Mike said. "It does for me. I think that's okay." But he's been wearing a prosthesis for forty years, he reminded me: "It would be different for someone else." Depending on how probing the questions become, he'll both indulge the fascination and explain the very real limitations that come with even this impressive, complicated object—a conversation that, in Mike's life, has become a ritualized social exchange set in motion by prosthetic parts.

Mike is an amputee, someone who lost a limb well after learning to navigate his world in a two-handed body. At fourteen, he got a diagnosis of epithelioid sarcoma in his left hand. Doctors had caught it early, but it required drastic action, at least in those days, because it's a cancer that easily metastasizes and spreads to other parts of the body. He and his parents made the difficult decision to amputate his arm just below the elbow joint. In the days that followed the surgery and recovery, Mike started the long process of occupational therapy to relearn manual activities, now outfitted with the "body-powered" model of prosthetic arm that was standard in its day and is still worn by many people in the world. It was a mechanical limb extension, anchored in a harness around the torso and tethered to the elbow, with a double-hook hand at the end that opened and closed only with shoulder shrugs and direct manipulation by the wearer; it had no added electronics to boost its capacities. The hooked ends wielded a bigger, blunter gesture than five fingers could usually accomplish, but it operated enough like an opposable thumb and finger that it could grip some objects in a relatively secure hold, with workarounds and adjustments.

Needless to say, adolescence wasn't the ideal time to acquire a suddenly conspicuous body, and Mike had to assimilate not just new tasks of carrying and gripping but a new identity, too. He resisted going to the beach near his hometown in those early months of recovery; it took time for him to internalize the "new normal" of the way he looked. He spent years with an undercurrent of worry that he'd never have a girlfriend. He weathered the unfortunate college nickname of "Clubber."

In Mike's kitchen that afternoon, I took a picture of his life's set of four prosthetic arms neatly lined up in a row—a catalog of the technology available to a middle-class person with ready access to modern healthcare. The prosthetic Mike wears most often now—the one that evokes all the comments at the dinner table—represents an impressive leap in engineering and a similar leap in expense: a myoelectric model, its motions directed by the electrical signals naturally built into the body's muscles, the ones that tell an ordinary flesh elbow how to bend or the wrist how to rotate. Myoelectrics create automated, more fluid motion for an artificial limb that results from this connection. They're designed to combine and amplify the range of movements a prosthesis can accomplish, ones that better replicate the fine-motor grasping or turning motions that are everyday moves for a flesh arm and hand but were laborious or impossible with previous generations of body-powered mechanical models. By the measure of technical novelty, myoelectrics are some of the best that money can buy—if you're looking for premium materials, a sleek, pristine shape, and some complex, finely tuned functionality.

Mike broadly enjoys the progression of technology that has brought him to the present day—he says he went, in other people's eyes, "from Captain Hook to the Bionic Man, overnight"—but the functional improvements that reentered his manual lexicon with the new technology are both less grand and more interesting than the average observer might imagine. He can now proceed, for example, through a buffet line by turning his prosthetic wrist to provide leverage for holding a plate while serving himself food with his flesh hand, in a relatively seamless inter-action. Attending a ribbon-cutting for the new wing of a local hospital, he found that he could clap his hands together in applause along with the other attendees for the first time since he was fourteen. That was a big deal—being able not only to reach or grasp or carry, but to participate in a social ritual, using the reciprocity of his palms in an everyday gesture that he'd only distantly remembered.

In the twenty-first century, the conversations that greet someone like Mike at the dinner table often contain the word *cyborg*, a catch-all term for human bodies that are aided or augmented by artificial parts. The word carries a variety of connotations, depending, Rorschach-style, on who's using it. *Cyborg* can signal a shorthand for the simple reality of contempo-rary life—flesh and automated systems operating together, from literal extensions like mobile phones to the more subtle automation created by birth control pills, in reciprocal relationships that make bodies and external parts nearly indistinguishable. Or the word can carry a more charged speculation about the future of human life with technology, functioning as an index of a person's relative optimism about the future of the body inter-twined with machines. Is the cyborg-self an ominous sign of a coming world under machine rule, or a promising fix for the frailty and idiosyn-crasy of skin and bones? Where does the machine start and the human stop? Does the distinction matter? These have long been favorite questions for enthusiasts of the posthuman, amateur and expert philosopher alike.

Donna Haraway, one of the cyborg's most notable theorists, named long ago what many of us unconsciously sense in the presence of high-tech tools: "Late twentieth-century machines have made thoroughly am-biguous the difference between natural and artificial, mind and body, self-developing and externally designed, and many other distinctions that used to apply to organisms and machines. Our machines are disturb-ingly lively, and we ourselves frighteningly inert." It's this *liveliness* of

Mike's latest high-tech arm that makes it such a draw, and the distur-
bance that his hybrid body brings along with it arises from an unspoken
but persistent fear: that a "smart" arm like Mike's, with its mechanical
cleverness, suggests a comparative inferiority of the human body, and by
implication, a fundamental passivity in the human being.

But talk about the body and its cyborg future tends toward speculation and even fantasy, and it can cloud the quietly extraordinary dynamism
of a body in its immediate world. Mike's prostheses tell a story that isn't
primarily about the machines; they are evidence of an endlessly plastic
and adaptive human body—human *person*—who survived illness in
adolescence and has moved through marriage and children and jobs in
adulthood, attended by these extensions as Mike found workarounds to
whatever situations arose. The arms are a partial snapshot of Mike's body,
the material culture of his back-and-forth relationship with the shapes of
his external world. That's why the term *assistive technology* is easy shorthand but ultimately both redundant and misleading—because assistance
is universal whenever we talk about tools. The way the body and machine work together is much more aptly expressed by another name for
prosthetics: *adaptive* technology. Tools don't run the show; they work
together with bodies in a mutual exchange of adaptation.

Even more than the futuristic talk of cyborgs and "bionic" prowess it
provokes, the outward appearance of the technology itself, with its persuasive cosmetic mimicry of body parts, can obscure how far even a
state-of-the-art prosthesis like Mike's lags behind the maneuvers that a
body is evolved to do. Mike gave me an example. To catch a football—an
oddly shaped projectile hurtling toward you, end over end—requires anticipating its uneven contours and its trajectory through the air, hands
and arms placed just so in preparation for its secure arrival. How many
micro-moves are included in this transaction? A dozen? More? Catching
a football with his prosthesis isn't in the near-term cards for Mike. It takes

a lot of time for deceptively simple tasks to become automatic. "For me to do this"—he reaches out to grab the seltzer can in front of him on the counter, a series of spatial moves he has learned to make, now impressively accomplished while holding my gaze—"that was about two years into the game." It took him five years to master stabilizing a fork.

Replicating the highly evolved mechanics of human flesh and bone is always difficult, even with the best available technology. It's especially difficult when trying to mimic anything approaching the range of motion in, say, the human wrist. Think of the complexities: a wrist will extend upward and down, but also side to side, plus any number of combined moves and motions for leverage, bracing, and flexing. It's an intricate mix of hard bone and soft tissue that works in an often unconscious feedback loop. Achieving that real-time responsive range of motion remains a fascinating challenge to engineers in the field, one of many in the quest for operational biomimicry. There's also the problem of getting the technology attached to the body without chafing or pain, and the pursuit of integrating prosthetic control systems more seamlessly with signals deep inside the brain. How do you build grip patterns for a set of mechanical fingers that will need to hold a lightbulb without shattering it and also to turn a heavy doorknob? All of these pursuits live in a domain of research commonly known now as Rehabilitation Engineering. The field is housed in the biomedical sciences, but its origins lie in war.

More than twenty thousand Union amputees returned from the Civil War, and prosthetics were made available to them for free. (To some veterans, that is; African Americans were barred from the veterans' associations and the postwar benefits that aided their white counterparts.) But many opted out, and it wasn't only because the devices were widely experienced as uncomfortable to wear. There was cultural meaning attached to the rejection of this choice, as a tribute song written in 1866 expresses well:

> *Three hearty cheers for those who lost*
> *An arm in Freedom's Fray*
> *And bear about an empty sleeve*
> *But a patriot's heart today.*

To "bear about" an empty sleeve meant to wear one's loss openly in public. Formal portraits of the era often show veterans sitting proudly in a uniform or civilian clothing with a pin prominently attaching the vacant wrist-end of a uniform jacket or shirt to the shoulder. A scar, a limp, or an empty sleeve became a way to announce a legacy of heroism. Putting loss up front told the story of a body these amputees could narrate for themselves—not the tale of swift replacement from old to new, but a story of prominent absence that spoke of valor. And of course this narrative was meant to ward off another possible interpretation: of bodies enfeebled or even destroyed by a prolonged and agonizing conflict, in which the loss of human life was unimaginably devastating.

But it was the vast number of veterans returning from the bloody wars of the twentieth century that provided the incentive to bring formal, sustained scientific and technological attention to "rehabilitation," in both a biological and a social sense. World War II, in particular, resulted in a greater number of men returning from the field alive, but with injuries that required amputations and therefore prosthetics and other assistive devices—in the United States, around twenty-seven thousand veterans in all, a number great enough to call for the formation of a new federally funded committee on prosthetic devices. In 1945, the National Academy of Sciences elected to devote significant new support to the research and development of prosthetics for returning veterans, with a noble rationale.

Providing replacement parts was not only a medical service to people who had lost some key mobility by injuries sustained in battle; it was also a way of restoring a foundational sense of *personhood*, a sense of wholeness following great sacrifice. Returning physical functionality, the thinking went, could mean allowing veterans to retrieve their former sense of identity—which could in turn include the capacity to hold productive civilian jobs in peacetime. Crucially, this population was distinguished from the broader population of the disabled. In general, the model of disability, then and now, was one of deficit: the body as subject to "handicaps" that rendered it weak and dependent. But in the case of wounded veterans, the sense was that they, as people—and the economic power they represented in the workforce—were being returned to a state of capacity they had once held.

Historian David Serlin writes that the stakes for rehabilitation were high, not just for the individual but for the country. A 1945 article in *Popular Science* told readers that injuries could now be "erased entirely or mended so subtly that no one knows of them except the men themselves and their families. . . . War's rehabilitation engineering may well become the *social engineering of the future*." It was ambitious claims like these, the "nationalist imperative" and "triumphalist rhetoric of Americanism," writes Serlin, that cast the idea of postwar medical care for veterans as "a social means to larger civic ends." The Servicemen's Readjustment Act, or GI Bill, passed in 1944, granted extensive coverage for prostheses and other needed equipment, but also customized structural remodeling for houses or cars—whatever was needed to return to life in a redefined "normal."

Rehabilitation Engineering provided these prosthetics and other gear as useful tools that were also metaphorical "passports," writes another historian, Henri-Jacques Stiker: a passport, a regenerative time-travel device to recover the old body and send it into a future with normalcy restored. This was recompense for personal sacrifice but also a reassurance to the country at large that its economic and patriarchal future would be secure, driven by men in familiar roles of rugged leadership, leading "ordinary" lives once more. Hollywood took up rehabilitation, too, with amputee veterans cast in starring roles to demonstrate their capacity both to tackle everyday tasks and to navigate (always heterosexual) relationships in their new bodies. It wasn't just the technology itself but the *stories* that mattered—the stories being written and rewritten on the bodies of these soldiers and for the nation in recovery that they represented. These artificial parts were making a powerful, expansive promise after the deep uncertainty of war: to construct, as Serlin calls it, a fully "replaceable you."

As a researcher and professor, I've spent a lot of time in the homes and workplaces of people like Mike, people whose prosthetic gear is technologically impressive, functionally meaningful, and compelling as examples of replacement parts. But once I understood the history better, I started to pay closer attention to bodies and people, whatever gear they use or don't use—to see people as the protagonists of adaptation in all its forms, taking up tools of all kinds. I got especially interested in tools that aren't attended by heroism and high-tech fanfare. Out of the limelight, for a fraction of the cost and with deceptively simple technology, there's a global material culture of the body reaching and grasping and supporting itself with objects that are all around us.

On an unseasonably chilly April night in the outskirts of Boston, I went to see Chris, who was trying to get a wriggling seven-month-old Felix to bed. Felix, olive-skinned and hazel-eyed, was fighting the emotional fragility of bedtime. In his small blue bedroom, he was easily captivated by the colorful board books spread across the rug. But he was also briefly and dramatically undone with desperate tears after bumping his head on the toy box, only to smile a moment later at the sound of his father's voice. This was just plain fatigue. Soon he'd be nursed by his mother, Melissa, and at last tucked in. But first: the changing table. Wrangling limbs and wiping chubby skin and swapping out a sodden mess for a clean diaper and then clothing this fidgety body in its tiny pajamas with sewn-in feet are among the last tasks of the evening with an infant, requiring a complex set of maneuvers without a lot of analogies in the nonbaby world. You can't really rehearse for it. It takes practice, in person.

For Chris, there was an extra complication. Chris has one arm and was born this way. He also has a slight Texas drawl, the compact and wiry body of a jockey, shoulder-length blond hair. He has the typical structure of a normative arm and hand on the left, and on the right, his shoulder extends to a mass of residual limb, skin over bone that tapers off a few inches beneath the joint. Eating, donning his clothes, and tying his shoes are tasks that most people think of as inherently two-handed. But Chris does these with some combination of his one hand, his two feet, his unusually dexterous toes—and, in key ways, his residual limb. And it would indeed be called a limb: it extends up and out as a gesture accompanying Chris's speech, and it casually hangs over the ladder-back of a chair. It braces surfaces like paper for writing and forms a clamp with his armpit. Stripped of clothing, it's a bit of a spectacle, not for its natural morphology but because of the ornate tattoo that encases it: the sprawling wings of a fierce red-eyed blackbird.

At the changing table, then, the job demands a series of planned movements, usually maximizing the work of two hands. Chris couldn't address this situation with his usual adaptive choreography of arm and limb. But he's an engineer, so he's trained to tinker. "At first I tried to think about fully automating the task," he said to me, hoisting Felix onto the table. A diaper-changing robot! But Melissa had wisely reminded him that diaper time wasn't just a task; it's also a moment of parent-child connection. So he went back to observing the two-handed method and tried to adapt it— how often a diaper changer holds the two little ankles in one hand while using the other to clean. With a bit of forethought and some controlled trial and error, he fabricated a low-tech tool from everyday materials: a holster, or perhaps more precisely, a rope sling, to suspend Felix's legs and backside so that all the clearing and wiping and swift replacement of wet-to-dry could take place. He showed me how it was done—"it's like a lasso concept," he said—hanging a loop of soft rope on the shoulder of his right limb, from which two C-shaped partial hoops covered with felted wool are suspended, with a clever knot above them to steady the legs, easily loosened and tightened as needed without pinching or chafing. Though it would never be recognized as such in the ordinary sense, this is a single-use prosthetic, an assistive technology that, for the cost of ten dollars or so, anticipates a baby's astonishing first year of constant growth and gets the job of diaper-changing safely done. It does the eternal work of the extended body, both extraordinary and utterly mundane. Both like and unlike Mike's arm, it's a tool for augmenting everyday tasks.

When Chris was six months old, a first prosthetic arm was recommended for him by his doctors—doctors with expertise that his parents, Vic and Barbara, still say they were grateful to have. Chris was their first child, and they had to make up the parenting for his atypical body as they went along. They were advised that Chris should learn to use the replacement limb as quickly and as early as possible. It was crucial, the doctors had said, to his development. Following that advice, they required him to wear the prosthetic limb for some set few hours of the day. In those years, they were uncertain how to weigh the expertise of medical practitioners against the evidence in front of them at home, when it wasn't always clear if the arm was providing assistance. "We tried to take our cues from him—if it seemed to bother him to have the prosthesis on, we took it off," Barbara told me. "We were learning as we were watching him."

Like any child, Chris was deeply adaptive. He remembers figuring out ways to play and explore without attachments or extensions. Without his prosthesis, he would scoop up two Lego bricks in his left hand, brace the first against his shoulder, and deftly snap the other on top, all by a series of one-handed moves that he was rapidly learning to execute. He both wore the prosthesis and resisted wearing it. His body was outpacing the purpose of the technology.

Even in those early days, Chris's parents noticed something more. Something was happening in the reactions to Chris wearing the prosthetic that went beyond an assessment of its functionality: "I realized a lot of this is whether you care about what it looks like," Barbara said. She could see that people were pleased to see him wearing the various arms that were made for him. There was less staring. But there were so many drawbacks. An artificial limb was heavy and hot, and Chris's body seemed so tiny to be lugging around that extra weight. It seemed to provide more interference than assistance: "There were so many things he just couldn't do with this big thing hanging."

What would this child need most from them—equipment in the form of medical tools? Or should they take a free-form wait-and-see approach to his dynamic young mind and body? Chris continued to make attempts to try the latest arm technology that was fashioned for him every few years, testing his habits and movements while they were still unfixed. With each new limb, he and his family would think, *Maybe this is the one*, the right fit to round out his functionality or perhaps to give back a cultural sense of normalcy to a young person whose identity was developing alongside his peers. Ultimately, he rejected each one. Instead, Chris began to develop various ways of answering other kids' questions about what had happened to his body: straightforward and patient, short and oblique, or, when he tired of the constant queries, sometimes just: "Alligator."

Chris became an engineer—not in prosthetics, as people often expected along the way, but in biomedical products. In a sense, he gave his life to technology. But he learned to trust that he would knit together his own adaptations to the world on the fly, using the body that he had as well as the physical features of the built environment that presented themselves, generally without specially designated assistive equipment.

At his kitchen counter earlier that evening, I watched Chris make dinner. "A lot of cooking is about steady surfaces, but everything comes down to the order of operations," he said, walking me through the steps. The sun was at a slant through the windows on all sides of a tiny dining alcove, an extending daylight that was a sign of spring. The baby bobbed and wiggled in a high chair, ready to eat.

Chris prepped an avocado on the cutting board, first by making one clean slice through the flesh to reach the pit. Next he clamped the knife handle, with the avocado attached to the blade, inside the armpit of his much shorter limb, and then grasped the avocado and turned it 360 degrees with his hand, slicing it in half against the stationary blade. From there he could separate the two halves, strike the pit lodged in one half once more, and clamp the knife handle again to pop out the pit, leaving two halves of the soft flesh. He does all this now without even thinking about it. To chop an onion for the soup, he'll execute a similarly internalized series of steps, a process that's been tested and refined and, finally, so fully rehearsed as to be natural, something Chris does while chatting about dinner and offering the baby some rice and beans, about half of which will make it to his tiny mouth, the remainder falling to the floor for the eagerly awaiting dog.

Each of Chris's moves depends on the shape and structure of the vegetable. As with the round avocado, he began by steadying the onion on the cutting board for the first cut through. Once he'd chopped it in half, he turned it flat side down for a series of stable one-handed cuts. He keeps his knives sharp for the cleanest slice. These methods use no prosthetic in any ordinary sense, unless you start to see the features of the everyday world—the relative friction of the woody avocado pit, the width of the knife blade and its grippy handle—suddenly alive with attributes available for steadying or pulling or other appropriation.

Coworkers and acquaintances often imagine that some situation or job will have to be laboriously adapted for him, but Chris often generates straightforward, elegant solutions. He'll use the sharp corner of a table to brace the stretchy wrist of a latex glove, holding it stable so he can insert his hand. Tying his hair into a ponytail is a lengthy series of moves not so unlike the two-handed version, the thumb and fingers of his one hand working the hair and the elastic against one another, but with much finer dexterity. Where opening the stubborn cap of a ketchup bottle might be a matter of brute force for someone with two hands, for Chris it's a subtler operation: he braces the heel of his hand against the bottle's neck and maneuvers the cap with that same hand's nimble thumb and forefinger, the whole task adapted to a more patient, more delicately articulated treatment. When he needs them, his toes, having been pressed into service from early childhood, can grip and manipulate with astonishing precision. With his left hand placed on a flat surface like a bed, for example, he can use his right big toe to clamp down on a set of clippers to trim his fingernails.

Any of a human's hundred daily motor-driven tasks—shampooing hair, or riding a bike, or opening a car door while carrying a heavy bag—involve a set of macro and micro movements honed until they are no longer conscious. They become automatic, internalized—but they can also be adjusted, depending on the job and the state of the body performing it. For Chris, tying his shoes or strapping the baby into his car seat involves adapting a whole series of interactions that were predicated on two-handedness; he turns them into a one-handed execution. The transposition earns an undeniable amazement when a two-handed person like me watches it unfold, but I've learned to temper my first response and understand this scenario as the magical-seeming evidence of something that's equal parts fascinating and entirely ordinary: a body adapting itself to the world, and the world to itself.

H ow would we compare these objects, Mike's high-tech arm and Chris's holster for changing diapers? High-tech or low-tech, one-off customizations or one-size-fits-most—it's hard to know just what to celebrate as adaptation, or innovation, or even what would count as "technology." Which replacement parts work best for which people? What tasks would be most important to any of us, if our bodies were to change tomorrow? And even if we could choose the so-called best objects, the ones that cost money and employ the most sophisticated machinery, it's not clear that those tools would perform the most vital work. Both Mike's and Chris's stories beg for us to return our attention to the *body* and the *person* as the site of infinite adaptation, but they also beg for an expanded definition of technology—not a simple contest of "better" and "best," but a broader canopy for how bodies meet the world of tools and environments for getting life done.

The historian David Edgerton, in his book *The Shock of the Old*, describes the lines along which much contemporary popular thinking about technology tends to run. Obsessions with the technological standouts of the twentieth century—flight, nuclear power, the birth control pill, the internet—are shaped by a myopic and linear view of history, one predicated on a belief in innovation that's realized in bursts of novelty, proceeding at an ever faster pace, and by definition "ahead" of the rest of culture. Many people tend to link technology almost exclusively with invention (the creation of a new idea) or innovation (the first use of a new idea), Edgerton tells us, focusing on research and development, patents, and early stages of implementation. And many of us tend to narrate the story of historical change brought about by technology along the same timelines.

But Edgerton shows how a history of technology-in-use yields a far different picture: everyday technologies like the rickshaw, corrugated iron, the sewing machine, and the bicycle have a long and profound arc of impact, but they get easily overlooked precisely for their workaday ubiquity. And "old" technologies can get new life over time. The condom, for instance, was assumed to be outmoded and replaced by the oral contraceptives that arrived in the 1960s, but the advent of human immunodeficiency virus (HIV) significantly revived its use, making it crucial in a new and urgent way to public health. "By thinking about the history of technology-in-use, a radically different picture of technology, and indeed of invention and innovation, becomes possible," Edgerton writes. "A whole invisible world of technology appears." Looking with new eyes at significance and utility, asking what counts as technology, gets us out of science fiction and squarely back into our everyday lives. Use-centered history, says Edgerton, "yields a global history, whereas an innovation-centered one, for all its claims to universality, is based on a very few places."

In a use-centered analysis of prosthetic technology, the standout would likely not be Mike's myoelectric arm or its high-tech competitors. Nor would it be the workarounds that Chris and others like him make for themselves every day. It would be a robust, low-tech and low-cost leg prosthesis that's being used every day by thousands of people all over India: the Jaipur Foot.

One of the largest public hospitals in Asia is in the city of Ahmedabad, in India's western state of Gujarat. The sheer magnitude and reach of the care it offers has resulted in an informal economy that thrives around and between its long row of buildings. Families camp out along its sidewalks and entryways, waiting for relatives or friends. Vendors in brilliantly colored clothing sell snacks, and dung patties for fuel, to these captive crowds, talking and trading as their paths are crisscrossed in every direction by mopeds, bicycles, animals, and pedestrians. And down at the end of the row, in the basement of a building that houses an extension of the hospital's offices, there's a small workshop for lower-leg prosthetics. This is one outpost of Jaipur Foot, a nonprofit organization that designs, builds, and distributes their eponymous artificial legs all over India and in surrounding countries, at hospitals and in mobile clinics to Asia, Africa, and parts of South America. The signature Jaipur Foot is a below-knee prosthesis, one designed to be the most robust and affordable of its kind.

On the day I stopped by the clinic, as a visiting professor running a design workshop at Ahmedabad University with my students in tow, a stone mason named Devansh from a tiny rural town was there for a fitting on what would be his fourth prosthetic leg from Jaipur Foot. Tall and taciturn, with a face weathered by many years of outdoor labor, Devansh told us the story of how he got his first foot. He'd gotten an infection after a bad fall twenty years earlier, and the infection had necessitated an amputation. He had spent a year without work while the leg healed, until he'd seen an advertisement for one of Jaipur Foot's mobile clinics on television: a team was coming to his region. Getting that first leg had made it possible for him to return to work, and each replacement has come when the prior one has worn out its functionality.

My students and I watched as Devansh sat with the clinic staff, who wrapped his knee in plaster that, once dried, would make a reliable mold of the uniquely organic shape where his limb ended. Jaipur Foot's limbs are made partly en masse and partly customized, in a smart mix of manufacturing and service that keeps costs low and distribution easy. The mold is used to shape the socket for more precise joinery between the leg and the plastic extension. The leg for Devansh wasn't made with metals or carbon fiber; there are no electronics or circuitry of any kind. It's made of a combination of rubber, lightweight willow wood, nylon cords, and high-performance polyethylene, a strong and waterproof plastic with joints for bending at the knee and ankle. These limbs do the crucial work of weight-bearing support and the bend-flex required for a walking gait. The rubber and plastic used in these models is heavy-duty, suited for multiple kinds of walking surfaces and weather conditions, and resilient over time; unlike prosthetic arms, the leg limbs have the benefit of gravity working in their favor, easily supporting upright human walking in a gait that presses down and pushes off for its locomotion. Each model costs around fifty dollars to produce, and most remarkable of all, almost all are given away for free, paid for by charitable organizations with local chapters. Since Jaipur Foot got its start in 1975, more than a million and a half of these limbs have been distributed in India and other countries, including places where land mines in conflict zones have created spikes in amputations.

I had traveled to India to see prosthetics like this one in use—the kind produced as technology for masses of people around the globe who are looking for replacement parts, a world away from the laboratories that create the elaborate customized "bionic" arms and legs in the domain of Rehabilitation Engineering, with its military-backed funding in search of simulacrum and enhancement. Jaipur Foot is a product and a service, built on networked locality and strong communications to assist people like Devansh, who might otherwise opt for a wheelchair in a country with very little of the hardscape that makes wheeled mobility possible. For him, the leg and foot were the difference between two decades of work and unemployment. The fitting on the day of our visit was brief and efficient. Soon Devansh would be on his way.

My students and I let all our questions multiply and then dwindle; we took pictures and shook hands repeatedly with the group of men who had assembled as our hosts. As my students left in ones and twos, making their way through the hospital campus maze and its crowds on their mo-peds, I thought about the product we'd just seen, inextricable from the web of connections that, together, produced the leg for Devansh. Jaipur Foot is just one of scores of examples like it around the world: organiza-tions using low-cost, readily available materials and local labor to create robust and elegantly designed prosthetics, suited to the living and work-ing conditions at hand, for those with little money to spare. But this very ingenuity brings with it a host of questions we didn't get into at the clinic. Would the availability of better healthcare have successfully provided treatment for Devansh's original infection and obviated the need for an amputation? What about the multidimensional irony by which global armed conflict creates the technical leaps that produce an arm like Mike's and also the land mines that inflict injuries that become amputations that necessitate prosthetic limbs?

With just a little bit of digging below the surface, you can find pros-thetics doing the work that all material culture does; they are artifacts that, under close attention, yield an index of infrastructure, local histo-ries, and social norms. They carry stories that precede their manufacture and follow from their user—into conditions of life, into economics and family and work structures and more, conditions that are partly inherited and partly chosen. Prosthetics, like other designed objects, are ideas made real in things. When they're used by human bodies, they become part of the story of those bodies, and the word *cyborg* could never even begin to indicate how deep and compelling these tales become. Cyborg talk is an easy passport out of the here and now and into a vaguely imag-ined future. Meanwhile, the parts-and-systems stories of everyday pros-thetics are infinitely more interesting. Anthropologist and prosthesis user Steven Kurzman is impatient with the term *cyborg* altogether, with its slick commercial appeal. A prosthesis is an extension of the body, not its driver; it's also a tiny node caught up in a constellation of manufacturing streams, politics, and history:

If I am to be [seen as] a cyborg, it is because my leg cost $11,000 and my HMO paid for it; because I had to get a job to get the health insurance; because I stand and walk with the irony that the materials and design of my leg are based in the same military technology which has blown the limbs off so many other young men; because the shock absorber in my foot was manufactured by a company that makes shock absorbers for bicycles and motor-cycles, and can be read as a product of the post–Cold War explo-sion of increasingly engineered sports equipment and prostheses; and because the man who built my leg struggles to hold onto his small business in a field rapidly becoming vertically integrated and corporatized. I am not a cyborg simply because I wear an artificial limb, nor is my limb autonomous.

Amputees (and other disabled people using assistive technol-ogy) are not half-human hybrids with semi-autonomous tech-nology; we are people.

A use-centered lens recasts the meaning of prosthetics when they land in our own lives, compelling us to think both about the material of the object and about the real wonder that's happening with all replacement parts: the wonder of human adaptation. What the gizmo does or doesn't do will always pale in comparison to the real miracle at hand—each body, endlessly plastic, responsive, operating with and through technol-ogies of all kinds to get its tasks done in a resourceful mix of workarounds, glitch-ridden patchworks, quick fixes, and slow evolutions. A more meaningful perspective on tools takes into account not only technologi-cal effects or how many units are bought and sold but the whole context of use and adaptation: when and how people opt in or opt out, access to supplies for repair, local customs, and always—*always*—who has the power to decide. Together, these parts and systems and the ideas behind them form a mixed-together story of how and when prosthetics arrive for and get taken up by people. It's what philosophers of science call the realm of the *biopolitical*.

P oet and essayist Audre Lorde, whose work spans topics of race, gender, and political rights, chronicled her treatment for breast cancer in the late 1970s in *The Cancer Journals*. Her book is full of meditations on what it was like to find herself a survivor of a terrifying illness. But the recovery process also brought with it all kinds of contradictory medical and cultural ideas about a normal, acceptable body, especially when it comes to women and their breasts. Lorde recounts the story of heading out to her postprocedure exam, the road ahead much on her mind. Her concerns were about her "chances for survival, the effects of a possibly shortened life upon [her] work and priorities. Could this cancer have been prevented, and what could I do in the future to prevent its recurrence?" she writes. "Would I be able to maintain the control over my life that I had always taken for granted?" Once at the doctor's office, however, it becomes clear that the choices before her carry meaning for others besides herself and involve more than her own recovery:

> Ten days after having my breast removed, I went to my doctor's office to have the stitches taken out. This was my first journey out since coming home from the hospital, and I was truly looking forward to it. A friend had washed my hair for me and it was black and shining, with my new grey hairs glistening in the sun. Color was starting to come back into my face and around my eyes. I wore the most opalescent of my moonstones, and a single floating bird dangling from my right ear in the name of grand asymmetry. With an African kente-cloth tunic and new leather boots, I knew I looked fine, with that brave new-born security of

a beautiful woman having come through a very hard time and being very glad to be alive.

I felt really good, within the limits of that grey mush that still persisted in my brain from the effects of the anesthesia.

When I walked into the doctor's office, I was really rather pleased with myself, all things considered, pleased with the way I felt, with my own flair, with my own style. The doctor's nurse, a charmingly bright and steady woman of about my own age who had always given me a feeling of quiet no-nonsense support on my other visits, called me into the examining room. On the way, she asked how I was feeling.

"Pretty good," I said, half-expecting her to make some comment about how good I looked.

"You're not wearing a prosthesis," she said, a little anxiously, and not at all like a question.

"No," I said, thrown off my guard for a minute. "It really doesn't feel right," referring to the lambswool puff given to me by the *Reach for Recovery* volunteer in the hospital.

Usually supportive and understanding, the nurse now looked at me urgently and disapprovingly as she told me that even if it didn't look exactly right it was "better than nothing," and that as soon as my stitches were out I could be fitted for a "real form."

"You will feel so much better with it on," she said. "And besides, we really like you to wear something, at least when you come in. Otherwise it's bad morale for the office."

Lorde understood the nurse's meaning immediately: that her choice about prosthetics told a story, and one that (rightly or wrongly) implicated more than just her. Her post-op prescription for prosthetics was never solely about functionality; it carried a social meaning, as had the doctors' recommendation that young Chris wear a replacement arm. Although it's harder to imagine such an exchange taking place today between a post-mastectomy patient and a nurse, a thriving cosmetic surgery industry speaks to the enduring influence of appearance on women (not only post-cancer but also postbirth, post-years-spent-on-the-planet). The emphasis on restoring the visual features of a body so that the loss and difference, the before and after, might be undetectable, can preclude the alternate possibilities that individual experience and proclivities might suggest.

It would be missing the point to read Lorde's experience as a blanket criticism of restorative surgeries and breast replacements. The choice to opt in to those technologies is personal and complicated, never a simple either/or; plenty of women in Lorde's position choose prostheses for their genuine advantages. This story isn't ultimately about the choice itself but instead presents an invitation—an opening—to see that an excessively narrow emphasis on the curative work of prosthetics can mask the equally urgent social negotiations that are always happening simultaneously. Lorde's book is a call for understanding prosthetic tools as a way to see gender at work, the roles that tools play in our cultural understandings of self. It came out in 1980, and it's safe to say that while things have changed, they have also stayed remarkably the same. A body is never the sum of parts that one chooses alone. Lorde knew this well, being a black woman, a lesbian, an activist, an all-around woman-out-loud in the 1970s. The stakes are high for people living with acute conditions of mismatch, ones that also depend on and change with their other markers of identity: race, gender, power. But bodies are *produced*, in part, by the material world around them. Prosthetics are political, as surely as they are biological.

A replaceable soldier, a bionic man, an injured worker, the ideal woman—these are stories about *people*, extended with prosthetic parts. The technologies that gave rise to them are functional, but they're also social, and the very meaning of disability is, too. The availability of and choice to use prosthetic objects tells a story about the wearer, and the evidence of their use, not their novelty, helps us understand their importance.

Bodies, power, and human worth—which kinds of tools will be the replacement parts that matter when our bodies change? Will we even know our own real wishes and our options when such a decision arises? I pay close attention to the moments when tools get their energy of necessity in the life of a person, because it's the will to reconsider, to reinvent and make new, that extends our reach. Replacements come in many forms.

At Cindy's house, there are dozens of cable ties around, some already in use and some waiting to be called into use. A cable tie is the humblest of objects—a strip of plastic a few inches long with a square hole in one end and a jagged, pointed end at the other. They're among the most flexible and affordable bits of hardware in existence.

Lots of people use cable ties to keep unwieldy things neat and organized; they're often handy as securing devices for the hidden infrastructure in a household or office. You can use one, say, to bundle a gangly set of electrical wires. Looping the pointed end through the square hole secures the tie with a ratchet-style hold, where the teeth of the end are tightly and elegantly caught by the edges of the square, preventing it from coming loose without any extra hardware needed. "I have dozens, if not hundreds of these," Cindy told me—cable ties prominently affixed to the drawer pulls on her antique dressers, through the heads of zippers on small bags holding change or lipstick, and to hang a parking pass on the rearview mirror in her car. She always has extras around, too, because—you never know.

Cindy is a quadruple amputee. At age sixty-three, she woke up to this new bodily condition, acquired after an extended coma following a heart attack she had while on vacation. After months in the hospital and hours of physical therapy, she returned home with her four limbs fundamentally changed. She lost both lower legs entirely and nearly all ten digits on her hands. She now has some broad grasping function on one side, but no complete fingers at all. It's a profoundly altered body she inhabits, one that calls for unusually deep adaptation. Cindy loves makeup and jewelry and lives in a suburban house outside Boston on a quiet leafy street, surrounded by dozens of photos of her children and grandchildren. She also has a wry and direct way of speaking about her life in this dramatic before-and-after: "I left for a week in Maine and came back to a completely different existence." And, to put a fine point on it: "I don't do change."

In the twenty-first century in the United States, people like Cindy, who are in possession of expensive commercial healthcare, become candidates for prosthetic limbs—the technologically sophisticated myoelectric kinds, like Mike's, that are fashioned to look like the human hand and, increasingly, to function something like it, too. Cindy calls hers a "Darth Vader" limb: it's black and shiny, with a zippered cuff for slipping on and off, made of machine parts that have been carefully crafted to articulate each digit. The hand has more than a dozen different grip patterns that can be programmed into the digits and activated on command— arrangements for the fingers and palm that allow for grasping a doorknob, for example, versus a wooden spoon. As with Mike's arm, there are sensors inside the housing that operate in conversation with the nerve endings of the body, speeding up the time between the impulse to perform a task and getting it done.

Cindy spent months working with her prosthetist on the insurance forms, and a week at the manufacturer's site in the Midwest getting trained to use this arm. She was nothing but hopeful, diligently stacking cones alongside the other amputees, eager to make friends with this replacement part and its promise of restored function. She'd heard the stories of others like her, for whom this arm was transformative in daily life. She kept her sense of humor intact, asking the engineers to form a grip pattern that included a solo middle finger pointed to the sky. This object, surely, was modern medicine at work, one bridge between her old and new body. It's the thing so many of us would imagine to be the ideal tool, if her situation were ours.

But at Cindy's house today, that arm is sitting idle in a closet. She ultimately found it hot, painfully heavy, and too beset with complications to use in any daily way. The arm that she worked so hard for turned out to be overkill, visually compelling and a stunning feat of engineering but a clunky behemoth in its design. In its attempt to mimic all handlike tasks, at least theoretically, it ended up fitting none precisely enough, at least for Cindy. So now the arm sits unused, while all over her house the cable ties perform their modest and necessary workaday utility. Instead of tuning a grip pattern for the myoelectric arm to open a drawer in her kitchen, Cindy uses the gross-motor grasping function she has to pull on the loop of cable tie attached to the handle, opening the drawer with relative ease and very little fuss or planning. A generously large loop of this lightweight plastic averts the need for the grasping fingers she no longer has. It's the right extension of the drawer pull that meets her body where it is, rather than a cumbersome attempt to restore her body to "normal" function that only succeeds in slowing her down.

Cindy's life is full of objects like these cable ties, a whole litany of low-tech tools forming a suite of prosthetics that would never warrant coverage in the shiny "innovation" pages of the tech magazines. Some of these objects are household products that are designed for the formal assistive technology market—a Velcro-secured hook for wrapping around a stick of deodorant, for example, making it easy to apply. And some are things Cindy has fashioned herself, born from a dissatisfaction with the available tools and a wish to regain some manual operations.

She was looking, for instance, to find a way to open a jar of cold cream or peanut butter, but she lacked the wrist-and-hand combination to do the task of turning. Having watched her occupational therapists do their nearly magical work of using everyday objects in new and adaptive ways, Cindy devised her own act of skillful assemblage. She took some peel-and-stick plastic hooks, the kinds that college students purchase for hanging posters in their dorm rooms without damaging the walls, and attached them to the tops of the jar lids. She can brace the jar with her right hand, and, with the residual parts of fingers on her left, she can grasp the hook with enough leverage to twist it open. Those extant fingers aren't nearly long enough to span the lid itself, but with the hook extension, Cindy can make it work. And the cost of each adaptation is under a dollar. What Cindy needs is not a single miraculous replacement limb but a whole panoply of these extensions, where the work is distributed among multiple objects chosen for the fine-motor calibrations needed to get the job done. Not a solitary "universal" arm, that is, but a series of objects that are just right and just in time. "Each thing I've found either by just cruising around on the internet or somebody telling me about it," she said. "I keep thinking that there must be other things out there that would make something easier for me, but I don't always know what it is."

The most striking object in Cindy's collection was fabricated for a few dollars at most, and it came not from her needs but from her wishes. When it became clear that the high-tech arm and hand weren't the solution she'd hoped for, she began to ask her prosthetist for ideas to recover the tasks she missed. One of those tasks was handwriting thank-you notes and letters to friends, a lifelong habit that Cindy prized. In an era of smartphones equipped with strong voice-to-text dictation software, she didn't *need* to handwrite anything ever again. But she *wanted* to, and this was the crucial matter. This specific attempt at replaceability carried meaning and purpose for her. She was making her difficult peace with her new body and its adaptations, and she was seeking the just-right tool to do it.

Her prosthetist team, a four-generation family business called United Prosthetics, run by the Martino siblings in the Dorchester neighborhood of Boston, started experimenting. This technical team—the tinkerers and builders of the world who operate outside the limelight of glitzy academic research—fashioned a simple silicone cap that was molded to the precise organic shape of her residual hand. It fit lightly but securely over the top and had a ballpoint pen lodged through two holes, at the precise angle for her to write. With some iterations and testing, and for a tiny fraction of the cost of her high-tech arm, Cindy and her prosthetist, Greig Martino, arrived at a bespoke prosthetic that now allows her to handwrite again—and in her own recognizable handwriting! Here was the extended body doing its work with tools, the combination of which is greater than the sum of its parts. She carries the pen cap everywhere in her bag for daily use: a prosthesis whose worth comes not primarily from the novelty of its engineering but from the close alignment between the self, a body, and one's desired interaction with the world. An object she herself was able to ask for, even without the technical expertise to fashion it. "After the pen came into my life," she said, "it seems such a simple design that I started to go crazy about: *What else could we do?* The obvious next step is, what else could we put in this cap that would make things easier for me? Eating is always something that you would like to make as easy and as inconspicuous as possible. Greig said: I can do a fork. I can do a knife. I can do a spoon."

If you assemble all of Cindy's everyday prosthetics today, you'd get a singular and eccentric archive. She has a sheet of foam core covered in nonslip plastic to hold her print newspaper open, and a hand-painted wooden tray, a gift from a friend, for holding her cards upright during her weekly bridge games. She's got a miniature set of tongs for grasping a sandwich, and the walker that provides support for her in-house mobility is outfitted with carabiners that hold all her supplies in what she calls her portable "command central." These objects together don't add up to a replacement limb; instead, in aggregate, they extend her reach.

Cindy's story is an invitation to think about the near universality of the changing body, especially with age. Plasticity works in every direction—the closures and openings that happen over time or all at once. A two-handed person who's just injured a wrist or broken a finger finds negotiating daily life to be elusive, at least at first. Faced with the sudden new barrier to our usual habits of movement, we might well ask: *Now how am I supposed to . . . make my dinner, take a shower, change the baby's sagging diaper?*

Even in these temporary states of change, it's easy to see only loss and diminishment, or to imagine that only narrowly defined replacement parts would answer our body's extended needs. But when the idea of wholesale replaceability reveals its limitations, what else is out there for the body in flux? Watching Cindy, and watching Mike and Chris, and summoning to mind those thousands of wounded soldiers, and arriving, finally, at Audre Lorde, expectant in the waiting room—each story makes it clear that the body is not only a human animal in a choreography with its environment. Every human creature's possibilities are produced, in part, by what's concretely present in its time and place. A body—any body—will take its cues, bend the available resources, and invent its being with the matter around it.

CHAIR.

From "do-it-yourself murder" to
cardboard furniture: Is a better world
designed one-for-all, or all-for-one?

The team at the Adaptive Design Association works with Niko to get the dimensions and features of his cardboard chair just right.

The Manhattan apartment where Mark and Vera live with their two-year-old son, Niko, is filled with light and with all the requisite toddler toys, crib, stroller, and other baby gear you might expect in a young family's home. In one corner of the living room there also sits a piece of furniture that's not part of the usual gear: a one-of-a-kind chair painted in two shiny tones of blue, its back and arms and seat mounted to a rectangular platform and complete with a detachable, soft fabric belt and a pillow to cradle Niko's head. Niko has looping brown curls and attentive eyes that fix on me regularly as he sits in this chair, tilted back to a precisely calibrated 30 degrees past the ordinary 90 of upright posture, a comfortable angle of recline.

I am talking to Niko's parents about a situation I understand, both in my work and in my life: what happens when the baby who arrives isn't the one you expected. Mark is a longtime clinical researcher in neurodevelopmental disease conditions, and so it happened that when they got some news about Niko just after birth, he understood its import immediately, more than most parents. "I knew what we were facing," he said. Niko, in his bright onesie, is coddled and cooed over while we talk; he is adored. And his diagnosis is a hard one.

Niko has a rare genetic condition called STXBP1, developmental and epileptic encephalopathy, which creates significant barriers to even everyday motor skills such as walking and speaking, and also causes complex cognitive delays. It's a rare genetic mutation that creates an atypical embodied life; Niko will always live with assistance of many kinds. At two, he is still in the early stages of building the strength to sit upright, but his parents and physical therapists know that it's best not to wait around. All play encourages both social and developmental skills in children—in parallel, not one at a time—and that combined play is what this chair makes possible. The seat and back are flexibly hinged to the platform in such a way that it allows Niko to sit at ninety degrees upright, to challenge the endurance of his trunk muscles, and it also tilts back through a range of degrees of recline, providing the close scaffolding he needs to play without frustration. When the chair is tilted back, Niko's weight is supported by its structure, and his trunk is held steady in a comfortable position without him having to use all his energy just to sit. The three of us sat and watched as Niko did just that, the secure posture freeing his arms and hands to work on fine-motor play and participate as a babbler in our conversation. On the big removable tray that attaches to the front of the chair, he practices the basics of toddlerhood, pushing the button on a plush toy, for instance, and seeing his action result in the toy lighting up. It's the practice he needs to press his fingers down with appropriate force, but it's also building his understanding of cause and effect, play that's both physical and cognitive at once.

Because it's been painted and sealed to be water resistant, you wouldn't know at first glance that this chair is made from cardboard, held together with dowels and tape and glue—the deceptively simple-looking form of cardboard known as triple-wall, each sheet of the thin fluted material fortified in a stack three layers thick. It's the kind of cardboard that the sturdiest moving boxes are made from, and the designers at the Adaptive Design Association (ADA), a storefront workshop in the garment district of Manhattan, have been using it to make low-cost one-of-a-kind furniture for people with disabilities, most of them children, all over greater New York City for three decades. You can find adaptive and assistive chairs in catalogs, of course, but this chair was made for Niko and no one else.

ADA is a two-story structure that's part office and part workshop, with a sunny window facing the midtown street that shows off the kinds of colorful chairs and other furniture that the staff there make. The display causes double takes among passersby in this section of Manhattan, which is mostly shops selling beads and buttons, parts by the hundreds. At ADA, each piece is custom made. On the days I visited, I saw dozens of chairs in all stages of fabrication—some in the early stages of manufacture, some getting taped and painted, and some getting ready to be delivered to their assigned house or school. There are all kinds of materials in the shop, but the most common is cardboard. It's everywhere, in big stacks of sheets and in small bits and pieces, getting sliced on the band saw and ripped through with box cutters, the spiky scraps collected in nearby bins. It didn't take me long to be romanced by all that this material can do.

Many people think of cardboard primarily as waste, but the way triple-wall is structured makes it surprisingly robust. On the one hand, it's only paper, the humblest of materials, ultralightweight and recyclable. But when three layers are stacked together, corrugated in between in a wavelike formation—what are called the "flutes"—its structure becomes improbably strong, holding a surprising 1,100 pounds per square inch, with the flutes acting like rebar in a building. Buttressed in key places, cardboard offers the right mix of precision and flexibility to let you build all kinds of structures. It can be cut with exactitude using a precision saw, but it's got enough pliability that you can score and bend it into shape or correct tiny flaws with relative ease. With taped edging and a polyurethane sealant, it can take a lot of wear.

Cardboard is also more than just a good physical material. It's a kind of working philosophical idea in concrete form. It's common to think of it just as a shipping container, an outer housing protecting valuable contents, or perhaps as first-draft material for an architectural model—a placeholder or rough sketch for the real thing to come. But cardboard has the affordances of both a draft and a finished product. It's strong enough to make furniture that is temporary, but also temporary-to-permanent, lasting years if needed. The material is sketchlike, under construction, with enough malleability to be edited, reshaped, or even discarded without too much preciousness. Cardboard has the virtue of being provisional, and it retains its experimental spirit even while it offers its sturdy strength. That unlikely-seeming combination of virtues—contingency and strength—is just right for fostering the design of adaptive furniture. It's also a great match for the near-magical plasticity that marks the development of young children, including Niko.

The machine shop and fabrication space is noisy and friendly, with a half dozen or so people working at a time, everyone in jeans and rolled-up sleeves, dust flying. There are power tools around the edges of the room, for fast cuts, and a CNC (computer-controlled) router and a 3D printer, because some jobs call for those tools. But there's also a stack of expired metro cards, perfect for spreading glue in an even swath. To score a line on cardboard for bending, they use the handle of a spoon instead of a conventional scoring tool. ADA could be called a *maker space*, to use the contemporary parlance, and the tools and materials that get the most use here, day in and day out, are box cutters, scissors, straightedge rulers, mallets, pencils (sharpened by hand on a wall-mounted model), an enormous amount of Elmer's glue, Velcro, and ordinary kitchen tools. This isn't a lack of sophistication or an excess of nostalgia for old-school techniques; it's a spirit of innovation made possible by friendly and accessible tools.

ADA's allegiance is not to any technique or material, but instead to keeping the entire process an open and democratic one, rather than to maintaining a protected place where expertise is cosseted and defended. If you're a parent or teacher with no building experience, the intention is to make you feel welcome here, too, because ADA believes that good ideas can come from anyone, not just designers, clinicians, and therapists. This isn't to say that anyone can do the fabrication work, which requires sophisticated skills. But the idea generation, the give-and-take over developing a model—in these aspects of the process, the staff are eager to say, anyone is welcome to participate. It's an ethics, almost a performance, this simplicity and humility, from builders who are as precise as any master carpenter. The workshop wants the message to be one of accessibility in the broadest sense: a reminder that the built environment and all its structures are the products of human decisions.

I've observed the ADA workshop for hours and days at a time over the last several years, especially the work of Alex Truesdell, ADA's cofounder and, until late 2019, its executive director. Truesdell is ADA's itinerant preacher. Whether teaching small groups about carpentry techniques, talking with families, or speaking to larger audiences in public, she is ADA's tireless evangelist-in-chief, the practical haircut and pastel cardigan belying her determination to alert everyone in her presence to the arbitrary design of the built world. "People think this is a furniture shop," she told me. "But really we're an explosives factory."

Women in "helping professions" such as special education are often patronized with bland praise for the "rewarding" nature of their work in a way that undercuts the radical proposition of what they do. Truesdell would know. She got her start as a teacher working with blind children in Maine. She began to conceive of a different path for her work when she witnessed her aunt's sudden loss of motor control in her hands after a spinal cord injury, and especially as she observed how her uncle responded to his wife's subsequent request for adaptations with ingenious, collaborative inventions. He fashioned leverlike extensions that allowed her to control round appliance knobs, for example, without needing to grasp and turn them, and he fitted a toothbrush with a tennis ball at its handle end so she could operate it without help.

Truesdell was both flabbergasted and frustrated by these solutions. The possibility for inventions in her own work lay right in front of her, she saw, ready to be cobbled together, but they also required not just mechanical ability but a mechanical *disposition*: the ability to see what was necessary and then proceed to build it, switching directions as trial and error required. Truesdell took an internship at Perkins School for the Blind, outside Boston, and there she began to imagine a way to serve Perkins students in this way: by making the exact thing that was called for, exactly at the moment it was needed, unfettered by the exorbitant costs of existing assistive devices and the long wait for paperwork to approve their purchase. Knowing that she would be hampered by her inexperience in fabrication, she began to teach herself basic woodshop skills by making things like high-contrast mobiles to provide visual stimulation for infants with low vision.

And then one day, when looking through an old storage closet at the school, she found a chair made of cardboard that had been built for a person with significantly asymmetrical leg and arm lengths. It was a light-bulb moment, she says. Here was a cheap and strong material that is particularly well suited to making semipermanent furniture for lots of bodies, but especially for the precise needs of bodies that are uniquely complex—the bodies of atypical children, which like all children's are rapidly changing.

Furniture became the focus of Truesdell's work to support children's learning and development. Instead of seeing a chair as a universal perch for childhood tasks, Truesdell says, a bespoke chair allows you to think about a singular child in their singular contexts: Does a child want to sit in a stable position, not at a table but instead on the floor, the better to play with siblings? Or reach a stand holding a musical instrument at school? Truesdell started to see ways to better sculpt the right tool for the job in front of her, customization driven by faith in the potential of the body, in the propulsive dynamism of any body meeting the world.

Very often that tool is a chair. I've never seen chairs more ingenious, more sturdy and compelling, more beautiful than the ones made at ADA, one bespoke and affordable model at a time. The workshop has created a chair to support a ten-year-old whose atypical legs weren't sufficiently strong for walking at all; she would climb into furniture and up and down stairs with her arms and hands, so they made a chair to anticipate that entry and exit. They make chairs that will hold oxygen tanks, chairs fitted for classrooms or cafeterias, chairs for work or play. They've made scooters with wide seats in lieu of standing platforms—using wood, PVC pipe, and wheels on casters low enough to the ground to move safely. They make footrests for standard chairs, and stools at every height. And sometimes, too, they design for unimaginable heartbreak: once, years ago, they assembled a lightweight carrier for a baby born with spinal muscular atrophy, for whom the experience of being held caused excruciating pain. Unable to hold her in their arms as they longed to, the parents went to ADA, who made the family a cradle-like structure from cardboard and padding. In it, the baby could recline comfortably but be held closely by her family for the eight months that she lived.

ADA has made customized rocking chairs, and supportive chairs for lying prone, and chair-table combinations to fit one child's needs and wishes precisely, down to a quarter of an inch. "Build for one," Truesdell told me, "and change everything."

In cultures where chairs are everyday fixtures of homes and offices, they seem timeless and permanent, as though they're necessary for survival. But the history of most furniture for sitting doesn't have much at all to do with what human bodies need, at least in terms of ergonomic structure and support. The chair is a revealing object that tells a story about bodies meeting the world in surprising ways—sometimes the cause of disability and sometimes its mitigation.

"Let's face the considerable evidence that all sitting is harmful," writes Galen Cranz, a design historian whose book *The Chair: Rethinking Culture, Body, and Design* traces this object's long global history. Not *all* sitting, of course. For bodies that use wheelchairs as mobility extensions, they're an elegant and crucial technology. And sitting itself is not the culprit; any unchanging, repetitive motion or posture doesn't give the body the variation it needs. But Cranz, writing primarily for an audience of ambulatory readers in industrialized and therefore sedentary cultures, is one of many researchers who've been announcing for decades that for many bodies, chairs are a major cause of pain and disability. Sitting for hours and hours can weaken your back and core muscles, pinch the nerves of your rear end, and constrain the flow of blood that your body needs for peak energy and attention. As even a brief survey of the industrialized world makes all too plain, most people's bodies are largely unsuited to extended periods in these structures in the present day, and never were in the past. Extensive research confirms that sitting in chairs is correlated, Cranz notes, with "back pain of all sorts, fatigue, varicose veins, stress, and problems with the diaphragm, circulation, digestion, and general body development." There's growing evidence that relentlessly sedentary jobs—in some, like bus driving and forklift operating, bodies are literally strapped to chairs—are harmful enough to shorten life expectancy.

For most of human history, a mix of postures was the norm for a body meeting the world. Squatting has been as natural a posture as sitting for daily tasks, and lying down was a conventional pose for eating in some ancient cultures. So why has sitting in chairs persisted and prevailed in so many modern cultures? As with all material objects, Cranz reminds us, function tells only part of the story. The other part, always, is culture— the inherited and sometimes arbitrary ways that things have always been done and therefore continue as common practice. "Biology, physiology, and anatomy have less to do with our chairs than pharaohs, kings, and executives," she writes.

One kind of historical chair, called the "klismos" by historians, evolved primarily as an historical expression of status and rank. Setting a body higher than and apart from other people, in an individual structure with rigid, flat planes—a throne, if you will—evolved as a way of recognizing an individual's power or leadership, with the earliest known models dating to ancient Egypt and southeastern Europe. Their use as an expression of authority continued through the Middle Ages and the Renaissance, and the endurance of this symbolism lives on as metaphor in many contemporary leadership titles; to chair the committee or the department, or to sit in the designated "director's chair" on a film set, is still to hold a seat of power.

In the centuries prior to Western industrialization, stools or benches were common household furnishings, but chairs were special-occasion objects, usually the exclusive property of the wealthy and powerful. The era of mass manufacturing in the nineteenth century, and the rapid social and economic changes that came with it, brought chairs into daily life for the first time. Industrial jobs, with their repetitive tasks, required a seated posture, and the high demand for chairs that this created in turn made them available and affordable to middle-class householders in Europe and the United States. "Chair-and-table culture," Cranz writes, has become fully entrenched in many parts of the world since then.

Modern interior designers have done their part to perpetuate chairs as a fashionable and practical norm, reinventing the form again and again in its aesthetics, though not nearly enough in its ergonomics. Domesticated in scale but striking in a room, chairs are four-legged creatures with anatomical backs and bottoms, familiar to humans because they stand up, almost like animals, beckoning us with their lifelike structures to sit down. Cranz notes that they appeal to humans, and perhaps especially designers, with this blend of the "architectonic and the anthropomorphic": they are structurally interesting and an echo of the body itself. But while they remind us of the human form, they rarely do much to actually support it. For instance, many chair designs come outfitted with the fleshy padding of cushions that seem to indicate comfort, but the consensus in ergonomics is otherwise. Cranz writes that "an overpadded chair forces the sit bones to rock in the padding rather than make contact with a stable surface, thereby forcing the flesh in the butt and thighs to bear weight." How can a nice cushy chair that screams comfort be so ill-suited to most actual bodies? The real science of ergonomics, Cranz argues, should point designers toward chair design that supports and enables the body's need for movement, not stillness—with seats that angle downward in front, for example, and have a base that's flexible enough for the sitter to shift their body weight from leg to leg. But for the most part, these principles are ignored in favor of fashion and cheap manufacturing. Chairs are generally not a response to the realities of the body, its natural evolution, or its needs for any extended period. Instead, the industrialized body has devolved in its needs and succumbed to chairs. "We design them," Cranz writes, repurposing a famous line of Winston Churchill's, "but once built, they shape us."

Naturally, there have been plenty of attempts by designers to reinvent sitting. There are kneeling chairs, bouncing balls, perch-style stools with rounded bottoms to encourage shifts in weight and movement. There are flexibly designed chairs like the Tripp Trapp for children, with pegs for adjusting the seat and leg supports to grow with a young body. Some office cultures have started to introduce standing desks. But at the average restaurant, in the ordinary classroom, and certainly in the subway car or airplane row, you'll still find chairs that are mostly at disastrous odds with any idea of comfort.

It's not just chairs, of course—so many of the products brought to market by the profession of industrial design weren't created for many bodies; they were designed to be plentiful, novel rather than necessary, and cheap. One designer memorably dubbed these bad designs a form of "do-it-yourself murder."

N ever before in history have grown men sat down and seriously designed electric hairbrushes, rhinestone-covered file boxes, and mink carpeting for bathrooms, and then drawn up elaborate plans to make and sell the gadgets to millions of people," wrote Victor Papanek in 1971:

> Today, industrial design has put murder on a mass production basis. By designing criminally unsafe automobiles that kill or maim nearly one million people around the world each year, by creating whole new species of permanent garbage to clutter up the landscape, and by choosing materials and processes that pollute the air we breathe, designers have become a dangerous breed.

Papanek was calling out his entire profession when he wrote these words. *Design for the Real World*, his first book, was a fiercely argued polemic about the misguided operations of industrial design in the mid-twentieth century, a clarion call for designers to question what Papanek witheringly referred to as *shroud design*—a preoccupation with the way things look on the outside, at the expense of how they should function and how robustly and sustainably they're made. He understood, that is, how the products with which we navigate daily life match (or don't match) the bodies and tasks they're meant to conjoin as tools. He likened his peers' shirking of their responsibilities to "what would happen if all medical doctors were to forsake general practice and surgery, and concentrate exclusively on dermatology and cosmetics."

A bestseller in its time, Papanek's book is now held up as a classic social wake-up call, joining a family of texts that includes Rachel

Carson's *Silent Spring* and Jane Jacobs's *The Death and Life of Great American Cities*—texts that are meant to rouse an industry or a culture from its slumber. Papanek was responding to the heady early days of truly mass manufacturing in the aftermath of World War II, when supply chains expanded to previously unimaginable scale, fueled by an equally behemoth advertising industry. Papanek understood the job of product design to be an ambitious one—to "transform man's environment and tools and, by extension, man himself"—but he saw in the giddiness of postwar growth a moment when designers lost their way. An overwhelming cultural belief and investment in technology created a cultural appetite for whatever seemed shiny and new, and the industry now pandered to it. The public's ready acceptance of these shiny new objects, whose life span was meant to last only until the next new thing arrived, created an allegiance among designers to what Papanek called "the dark twins of styling and obsolescence." Manufacturing for styling or shroud design, amplified by advertising, resulted in an empty desire for unnecessary objects that became quickly unusable. And objects that were created purely for ephemeral desire, in turn, created a whole ethos of obsolescence—an accepted disposability that made for a dangerous neglect of safety standards, resulting in needless injury from common household objects. At the time of his writing, Papanek claimed that six hundred women annually lost a hand in injuries brought about by top-loading washing machines. Eight thousand, he wrote, had died using fire escapes—objects whose very reason for being was to provide safety.

When operating at its worst, design resulted in needless injury, but in its everyday mediocrity it also created rampant misfit conditions. Here is Papanek's irony-laden takedown of a typical design process, this one about our subject of chairs:

> A new secretarial chair, for instance, is designed because a furniture manufacturer feels that there may be a profit in putting a new

chair on the market. The design staff is then told that a new chair is needed, and what particular price structure it should fit into. At this stage, ergonomics (or human factors design) is practiced and the designers consult the libraries of vital measurements in the field. Unfortunately, most secretaries in the United States are female, and most human factors design data, also unfortunately, are based on white males between the ages of eighteen and twenty-five . . . the data have been gathered almost entirely from draftees inducted into the [military branches] . . . there [are almost] no data concerning real measurements and statistics of women, children, the elderly, babies, the deformed [*sic*], etc. . . .

When we work as a cross-disciplinary team to design a better chair for secretaries, who are we designing for? Certainly the manufacturer wants to build secretarial chairs only to sell them and make money. The secretary herself must be part of our team. And when the chair is finished, there must be real testing! Nowadays an "average" secretary is usually asked to sit in the new chair, sometimes even for five minutes, and then asked, "Well, what do you think?" When she replies, "Gee, the red upholstery is real different!" we take this for assent and go into mass production. But typing involves eight hours a day, long stretches of work. And even if we test secretaries intelligently on these chairs, how can we see to it that it is the secretaries themselves who make the decision as to which chair is bought?

This scenario will be familiar to many people even fifty years after Papanek's book was published. There's a wide gulf between the people who use TV remotes, car dashboards, and other everyday items and the people who make the design decisions that bring those items into being, who are driven or constrained by motives other than ease of use. The *bodiedness* of people gets lost, especially in the sedentary workplace dotted with scores of chairs, where human labor is imagined not as flesh and muscles but as measurable economic deliverables, organized by roles and tasks. Yet the work is carried out by bodies all the same, not just in chairs but among the angular shapes of desks and copiers and countertops, alongside mechanical and digital machines. The passive as well as active physical requirements of our workdays tax our bodies, whether we bend repeatedly to harvest berries or sit still in a phone bank for hours on end.

Papanek was particularly barbed in his assessment of design's failure to allow for non-normative bodies. He called for a much closer focus on people and conditions whose needs were commonly written off by his design counterparts as "special": older people, people with disabilities, elementary school students, and any population considered beyond the scope of the middling mainstream. But as Papanek points out, we were all children once, and almost all of us become, in turn, adolescents, middle-aged people, and older adults. If we combine all the "seemingly little minorities" and their "'special' needs," he wrote, we discover that "we have designed for the majority after all."

Papanek was so nearly there in his recognition of the prevalence of disability, and in his insight that it is ordinary people—the users themselves—who are the experts on the shortcomings of industrial designers, even though his use of "the deformed" to describe disability speaks volumes about the limitations of his cultural moment and his own awareness. Indeed, disabled people have understood "murderous" design since long before Papanek's time and have acted as informal and formal designers or collaborators, often operating outside the public eye, to reshape the status quo of the built world.

P ost-Papanek, designers continued to wrestle with the conundrum of chairs. In the late 1970s, the office furniture giant Herman Miller commissioned independent designers Bill Stumpf and Don Chadwick to look for opportunities in the furniture market, gaps that new products might fill, and thereby expand the company's market share. In the course of their research, Stumpf and Chadwick, seasoned observers of humans and designed environments, noted one particular context in which chairs were creatively adapted but ill-suited to their sitters: older adults who spent a lot of their day sitting in La-Z-Boy recliners, watching TV or doing otherwise passive activities. They noted that the recliner had even become a makeshift medical treatment chair; they witnessed it being used, for example, in a half-reclined state for people undergoing dialysis treatment. Despite the advantages of its variability in pitch and support structures, however, it was the *wrong* chair for many of the conditions that come with aging. Older adults with weakened muscles had trouble getting in and out or reaching the lever for changing positions, and the deep padding, used for its presumed comfort, put people who sat for long periods at risk of bedsores.

Inspired by the spirit of the recliner, Stumpf and Chadwick reconsidered all of its traditional features and generated a prototype called the *Sarah chair* that they pitched to Herman Miller in 1988. It had all the flexibility of a recliner and then some—multiple ways to choose the pitch of the back, seat, and arms, with a flexible footrest for changing positions more easily. They reduced the padding significantly, making a thinner and therefore more breathable seat structure. The model was popular in-house as a chair with purpose—many Herman Miller staff could imagine it being useful for their grandparents, for example. But the mass-market appeal wasn't obvious at first, and the firm rejected the Sarah in its first iteration, only to reopen the design a few years later.

For the second version, Stumpf and Chadwick imagined the ergonomic features of the chair with aging still in mind, but also the body of the average office worker, commonly tethered for long stretches to a personal computer and keyboard. They got rid of the foam padding altogether, making the model's signature and surprising new aesthetic a cushionless seat of thick plastic webbing that would support the shapes of any body, not just an aging one. The Aeron was a wild success, quickly taken up as a status symbol in high-end office environments like those in Silicon Valley.

Aeron chairs have become a canonical, if expensive, example of what's called "universal" design and its variants, barrier-free and inclusive design. This principle is what many people think of when they hear the words *disability* and *design* together. The logic of universalism goes like this: designers gain surprising and powerful insights from looking closely not at norms and averages—not at people whose experiences fall well inside the expected middle of a curve—but instead at people and scenarios at the margins of experience, so-called extreme users. Good designers, the thinking goes, will take a close look at unusual circumstances, places where products (or environments, or services) are full of friction—barriers or ill fits or inefficiencies—for people with particular needs. There, in the margins of human experience, are clues to suboptimal conditions that may also affect people in the normative middle, though perhaps to a lesser degree.

In the case of disability, that would mean closely observing older adults and wheelchair users and people on the autism spectrum, say, and then abstracting from those observations, aggregating those clues for insight. What's working for these people, and what isn't working, based on what they say or do, or what they *don't* say or do? Where might improvements—or *interventions*, as they're often called—be introduced to address those points of friction at the margins, places where design isn't working in the extreme cases? After a series of prototypes and testing, the process in theory yields a more user-friendly, ergonomic, accessible product (or service, or environment) for all—in principle at least, design "for everyone."

It wasn't Papanek who offered universal design to the world; it was disabled people whose long-held insights generated what became a set of principles designers could follow. Ronald Mace, a wheelchair user who became an architect devoted to the theory and practice of accessible architecture, is credited with introducing the term *universal design* to the public in 1985. In part, the coinage was strategic, recasting features of design that had been considered "special" as simply good design, resulting in products and buildings that were straightforwardly "usable by all people." The universal design principles that Mace and others in the disability community generated read like an antidote not only to the preoccupation with "styling and obsolescence" that Papanek called out but also to the "murderous" results of negligent design; they include principles like "simple and intuitive use," "perceptible information," "low physical effort," and perhaps most important of all, "tolerance for error."

Universal design isn't restricted to high-end items like the Aeron chair. If you go looking for kitchen tools in your average big-box store, for example, you'll easily find the OXO Good Grips line, with the familiar thick black pliable rubber handles that have just enough give to hold comfortably and just enough sturdy resistance to offer leverage. The brand began with a vegetable peeler, an object so everyday as to be invisible, and the peeler originated, as design often does, in a complaint that inspired an idea.

In the late 1980s, a woman named Betsey Farber was attempting to use a standard-issue metal vegetable peeler in the vacation house she was renting with her husband, Sam, a retired entrepreneur. The tool was frustrating to use, especially because Betsey had arthritis. The misfit sent her speculating about what would make a better tool and hand tools like it. Right there in the kitchen, Sam and Betsey started to sketch out ideas for a new version of a peeler. With further refinement, those ideas resulted in the now-ubiquitous OXO design: fins on the sides elegantly indicate the place to put your thumb to optimize the mechanics of use, a visual cue for intuitive handling. When you hold the peeler against a carrot, it encourages you to apply just enough pressure to catch a ribbon of skin without

WHAT CAN A BODY DO?

cutting too deeply, and without a reckless amount of slip that would send both carrot and peeler flying from your hands. Grip and friction and leverage and force—all of these considerations have to come together in a quick intersection of concerns. The properties of even an object as simple as a peeler become consequential when it's used every day, as peelers often are.

Sam Farber came out of retirement and went to work on the peeler, and the idea that he and Betsey hatched was taken up in a partnership with a product design firm that went on to create the Good Grips line: can openers, salad tongs, and other tools made for the subtle work of manual kitchen tasks. Thus went a classic universal design success story, which has taken its place in the canon alongside that of the Aeron chair, as inspiration to a generation of would-be practitioners. There are other stories besides.

The Fiskars scissor company designed a set of shears they originally called their Golden Age model, built with a spring to open the blades freely and automatically, and with handles that rest on a surface while cutting—features that made their use more amenable to the hands of older adults or others with less manual dexterity. But when the company tested the scissors with focus groups, they found that the general response was strong. So they renamed and marketed the scissors as the Softouch model for wider appeal.

Designer Marc Harrison was commissioned by Cuisinart in the late 1970s to overhaul its commercial-grade food processor into a more consumer-friendly domestic model. The company didn't ask Harrison to address disability as a use case for its product, but Harrison's research and teaching at Rhode Island School of Design had included years of collaborative work on furniture for rehabilitation settings and wheelchair-accessible housing. Those experiences nourished the insights Harrison brought to the processor. He introduced things like large-print, high-contrast lettering for the labels and controls that flipped like paddles rather than requiring fiddly turning. These and other features were informed by usability principles Harrison had learned from looking closely at disability. This kind of commitment and imagination in design yields products and experiences that can make life better—without your even noticing it.

Disability is the little-acknowledged heart of the innovation in many digital tools, too. The history of the telephone, for example, is tightly bound up with research on deafness. Alexander Graham Bell's work with deaf students and their teachers solidified his research on making speech visible, reducible to signals, and therefore transmittable by electronic means. That research made telephone technology possible, but it also helped set in motion the standardization of signal processing that was crucial to early computing. And perhaps most hidden-in-plain-sight of all: consider the landmark change to media consumption that happened after disability activists fought for closed-captioning technology to be built into every standard television. The Television Decoder Circuitry Act, passed in the United States in 1990, required the technology to be a standard part of the television manufacturing process rather than an extra device attached alongside it. The battle to pass the legislation was difficult and protracted, in the face of resistance from within the communications industry about the perceived extra expense it would entail. But thanks to the economy of scale created by the legislation, the cost of adding this capacity to televisions is now so small it's almost impossible to calculate. Meanwhile, closed captioning has become a standard feature of daily life. It's how you follow along with the championship game across a restaurant or airport terminal, and it's how you understand clips from an election debate on your laptop with the sound off, making it possible to multitask when your kids are in the next room trying to get to sleep.

It would be easy to conclude from these stories that universal design is dispatching "murderous" objects to the dustbin of history, and that it alone points the way to a more accessible future. Designing ergonomic objects that can be mass-produced and affordable enough to find their way into the hands of people who need them is an unassailable good. But disability scholars also point to the tradeoffs—wins and losses—created by the work that Mace and his ilk set in motion. One is that while the dominant model of universal design has disability at its center, the very success of the innovations it generates tends to obscure their origin stories, as in the case of the OXO peeler. That success erases the very real, ongoing barriers to an adaptive, flexible world for disabled people. Universal design also tends to stoke an unquestioned faith in the importance of *products*, attained by *consumers*, as the key to building a desirable world. A "better mousetrap" may mitigate barriers in the short run, but sometimes it's a better process, or a better system, that's called for to lessen the conditions of misfitting meaningfully.

Take, for example, the Leveraged Freedom Chair, a design created with paraplegic wheelchair users in parts of the world where accessible streets are rare. Amos Winter, director of the Global Engineering and Research Laboratory at the Massachusetts Institute of Technology (MIT), did years of trial-and-error field research with his team in Tanzania, Guatemala, and India to collaboratively design a chair with variable torque in driving, even in off-road terrain, using very simple technology. Depending on where you grab one of its extended handles, you get more or less mechanical advantage, just like switching gears on a bicycle, which makes uneven pavement or dirt roads much more navigable. A third wheel in front adds balance. "We failed a number of times," Winter has openly said. The design process "has to start and end with end users," he says. "These are the people who have to define the requirements of the technology, and give it the thumbs-up at the end."

Getting the seat, frame, and handles in the right positions to be both comfortable and advantageous in their physics was an engineer's task of very fine designing and building. But the real ingenuity isn't in the mechanics; it's in the sustainable *system* for repair and maintenance. Many wheelchair companies make their chairs with their own proprietary parts, so even if they donate free chairs to people who need them, the parts are expensive to replace and the chairs become unusable when they break. If a wheelchair is going to last in a rural area among populations who aren't rich, says Winter, "it has to be repairable using the local tools, materials, and knowledge in those contexts." The Freedom Chair was made of globally standardized and affordable bicycle parts, so they can be repaired and replaced anywhere in the world. "To make something cheap, and simple, and reliable often takes rigorous engineering," says Winter. "What is the solution that will give you the required performance for as little money and as little complexity as possible?" The Leveraged Freedom Chair was not a compromise; it wasn't a chair that was "good enough" for the people who use it. It was the right design for specific people in specific contexts, with deep attention to longevity built into the model.

Parts and systems, biological and cultural, universalized or sustainably just right—as a researcher, I've long collected stories like these, of chairs and other household products where the redress for misfitting says something big about bodies meeting the world. But the cardboard chairs at ADA have stayed in my mind, especially for the ways they turn the idea of universal design upside down. Chairs and other furniture are designed at ADA for hundreds of children, one bespoke and affordable model at a time, each an instance in which universalism reveals its limitations and each an invitation to think about mass production differently.

Back in their living room, Vera and Mark told me the origin story of Niko's chair. Anyone can initiate a design-build request and work with ADA to get a custom piece of furniture. In their son's case, it was his physical therapist, Yasmin. At around age one, the developmental goal of sitting up was still elusive for Niko, so Yasmin suggested that the family partner with ADA for a chair that would support the goal. After an initial meeting to talk over the possibilities, there was a first fitting, in which Vera and Yasmin and the ADA team met to try out an initial prototype with Niko. The fitting is meant to address whatever flaws arise in the actual model in progress, not just in what can be surmised from drawings and measurements. Niko sat in the chair atop a large work table, and the team, clamps and measuring tape ready, watched his face for signs of discomfort or irritation while they checked the physical dimensions and his posture. The team made some tweaks accordingly: the V shape of the headrest wasn't quite right, for example, so the team decided on a wider U shape that would offer him a more comfortable alternative. After a second fitting, the team delivered the final chair, the one I saw that day at their apartment. It's built with some adjustability to anticipate how Niko will grow. It has two tray attachments and a seat insert to lift him higher from them when he grows taller. A modular footrest at the base has a layer that can easily be removed as his legs grow longer.

In other words, the chair is not only bespoke but also malleable and adjustable, just like the child who's in it. Vera and Mark are raising a baby for the first time, so they have no comparison experience for the rate at which his skills will come. Just asking for the chair marked a moment of reckoning—"a moment where you admit that your child needs help," as Vera put it. On social media, where parents of disabled children find each other and provide mutual support, Vera had picked up a way to think about the pace and quality of Niko's growth that would be evident, in part, from his sessions in the chair. Families of kids with STXBP1 post updates about their children's "'inchstones' rather than milestones," she told me. It's a way to signal the different steps in the trajectory they celebrate—and they do celebrate every one. "It's not just that he's going to be slower," she said. "It's more than that." Niko's path, and that of his parents, would be one forged each day with a lot of unknowns. It would be one of a kind.

Niko sat, alternately babbling and listening, while we three perched on the floor nearby as the afternoon wore on. Watching young children play like this is always mesmerizing to adults in the room, all eyes trained on the tiny person—a friend of mine calls it "Baby TV." But given the substantial complexities of his body and the odds of the life in front of him, most people would assume a child like Niko possesses no authentic toddlerhood to speak of. They can see the diagnosis and nothing else. Mark and Vera are very much aware of the realistic contours of their son's future life, and by implication, their own. But as we sat there in the sun, watching Niko, speaking questions addressed to him and answering them back to ourselves in the irresistible high vocal register that babies love, I thought about all that this chair stood for. The structure was so modest, but it carried in its material the power of belief: that a child's dignity is absolute, that he is present, remarkable as he is, singular in gifts—that every mind and body is plastic and curious. That he is, above all, loved.

Who was the designer of Niko's chair? The process through which it was created is an instance of what's called *co-design*: the ideas belong to the group, and the end product is the result of an interactive social process. It's also an instance of what design researcher Ezio Manzini calls "diffuse design"—where the work is "put into play by non-experts with natural designing capacity," a companion and contrast to the "expert design" that's been the industry-driven expectation for how the built world gets its shape and form. Humans have always had a role in fashioning and refashioning their individual worlds, of course, but the modern demand for mass-manufactured consumer goods created and elevated the role of industrial designers as experts in delivering many of the everyday tools and gear that humans use. Diffuse design requires a broader definition of expertise: Niko's parents and therapist needed the technical expertise of ADA staff, and ADA staff in turn needed the observations and preferences of Niko's parents and therapist—and the reactions of Niko himself, young as he is—to get the structure to Niko's liking and needs.

It's not the specific materials or techniques that are used that matter, Manzini emphasizes, so much as the collaborative associations that are formed when universal challenges are considered in a local and unusual way: the agency that comes from reformulating a problem. At ADA the quest is not for the perfect universal chair but a disposition toward a sustainable and collective form of working on adaptive tools, one at a time—for many. When it comes to truly urgent questions, Manzini writes, "radical innovations generate answers that change the questions themselves."

That's the challenge to universalism. Is the all-for-one, one-for-all model the only approach to universal design on offer? Niko's chair suggests an alternative model: local collaboration on the specific interpretation of a universal idea—chair—relying on low-tech materials that are available almost anywhere. And this additional layer: shared ideas and information, both digitally and face-to-face. ADA's website features dozens of resources for replicating its work, including a free video library for sharing tools and techniques. The group also offers short- and longer-term trainings, either at their site in New York or in intensive workshops in Haiti, the United Kingdom, Ecuador, Argentina, Kenya, and elsewhere. It's the *method* that makes ADA scalable, not its products. Manzini terms this way of working "cosmopolitan localism"—where "places are no longer isolated entities, but rather nodes in both short- and long-distance networks," connecting "a particular community to the rest of the world." The concept could apply equally to community-supported agriculture systems in rural China or groups of older adults in Italy who offer local college students an affordable place to live in their homes, in exchange for household assistance, as a creative approach to "aging in place." Cosmopolitan localism unites small practices and communities to a connected, globalized network for sharing ideas.

A small clinic in Ecuador gives a good illustration of how ADA's approach to process rather than any particular widget is what most fruitfully scales in a global context. At Prótesis Imbabura in Ibarra, as at rehabilitation clinics all over the world, post-injury rehabilitation services and prosthetics are designed for people with disabilities. But in Ibarra, where financial resources are scarce, for the last ten years clinicians have been building ingenious cardboard adaptations. In addition to furniture, they've built cardboard supports, covered in soft padding and fabric, to bridge the gaps between children's bodies and adult-scale wheelchairs, which are often the only wheelchairs available. They've done the same with standard strollers, which sometimes serve as adaptive "wheelchairs" for disabled children even up to age twelve. Strollers are far easier to maneuver onto buses, because they collapse easily; cardboard props can give needed rigidity to footrests or fabric sides. Like the Leveraged Freedom Chair, but in parts and pieces rather than in the form of whole objects, these adaptations are not distant echoes of their high-tech counterparts. They are the nimble, lightweight, convenient "just right" tools at the right time. They provide diffuse, elegant solutions for the precise wishes of the people who will use them every day.

Adaptive or diffuse design, at ADA and elsewhere, isn't meant to substitute for standardized devices or the engineering that produces them. Disabled and nondisabled people will always have needs for these devices—wheelchairs, both motorized and manual; eyeglasses; hearing aids; and more—and will continue to use them (if and when they can afford them). Rather, its focus is on augmentation and alteration—on the entire *ecology* that is required to make the world more meaningfully accessible, especially when a quarter of an inch makes all the difference. The tools and materials aren't just affordable and commonplace; they offer the invitation toward a make-it-yourself disposition and belief that the right modification is out there, within the imaginative grasp of not just experts but networked collaborations of ordinary people. In the language of manufacturing, it's sometimes called mass customization.

Still—most people hearing the story of cardboard carpentry insist on asking: *Can it scale?* This question is everywhere not only in the world of design but in nearly every aspect of contemporary life, from business to education to housing. When I hear it, I always think of a famously instructive story recounted in *The End of Average* by Todd Rose, director of the Mind, Brain, and Education program at Harvard University. Alarmed at a high incidence of crashes during routine flight training in the 1940s, U.S. Air Force officials looked for evidence of mechanical flaws in the planes or human error perhaps inadvertently introduced by their curriculum, but the cause of the crashes remained mysterious. At last, officials commissioned a lieutenant trained as a scientist, Gilbert Daniels, to look at the physical structures of the cockpit and the men who used them.

Daniels noted that all the cockpit structures—seat and back, pedals, knobs, and so on—had been built to specifications calculated for an *average* military recruit. Recruits for pilot training were already selected for some degree of averageness, had been the reasoning, so these dimensions should fit most pilots, most of the time. But when Daniels measured 4,063 soldiers, he was astonished to find that not a single one of the men fit all ten of the measurements that had been determined to be average. Instead, every body offered its own variation:

> One pilot might have a longer-than-average arm length, but a shorter-than-average leg length. Another pilot might have a big chest but small hips. Even more astonishing, Daniels discovered that if you picked just *three* of the ten dimensions of size—say, neck circumference, thigh circumference, and wrist circumference—less than 3.5 percent of pilots would be average sized on all three dimensions. Daniels's findings were clear and incontrovertible. *There was no such thing as an average pilot.*

The unyielding fixity of the average cockpit ended up being useful to exactly no one. Thereafter, aeronautical engineers began to make everything from seats and foot pedals to flight suits and helmet straps adjustable, and the Air Force adjusted its cockpit specifications to stipulate movable parts that could be adapted to fit a range of body measurements, from 5 to 95 percent of average, just right.

If you visit the ADA offices, you'll see movable parts everywhere, not just for their child clients but for everyone who works there. At least half of the furniture in the office is nothing you could buy in a store. Everyone at ADA sits on cardboard stools and perches—unless they're also wheelchair users, in which case they may have other extensions or structures around their desks to make their work more adaptive to their bodies and gear. Everyone there uses furniture that's more ergonomic than anything you could purchase at an office supply warehouse; every body takes advantage of the opportunity to be custom-fit.

It's a world under construction that extends from the furniture to the labor model itself. Since its inception, ADA has run in part by employing women in need of a work placement as part of New York State's Alternatives to Incarceration program. These women learn cardboard carpentry and collaborative design, tools and techniques; they are also refashioning their own lives with solid job experience, one move at a time.

The adaptive approach—the stance, if you will—is infectious, consciousness-changing. As I read Cranz's book and looked through the notes of my visits to ADA, I noticed myself straightening my curving back, becoming newly aware of the weakness in my spine and stomach, which did not seem strong enough to hold my body erect and stable for any length of time. This was doubtless partly from lack of exercise, but also from being in a body industrially, repetitively devolved to depend on chairs. I recalled the first time I went to ADA, where I sat on one of their cardboard perches, pitched precisely at the angle recommended by researchers and ignored by most schools and workplaces everywhere. I had felt my spine lightly snap naturally to its most supported position— straight up and relaxed, but without effort. To truly be with ADA is to see the alteration of chairs and other furniture with fresh eyes for history—to see alternatives to chairs and other furniture so pervasive they have become invisible.

And then I thought about Niko in a chair of his own, with an adjustable structure and measurements made just right for a child who will never be an Air Force recruit, and who may only express his preferences in a language other than verbal speech. At ADA, a person doesn't have to pass a fitness test to be worthy of a chair that precisely suits them. Do any of us, anywhere, I wondered? I thought about Adolphe Quetelet and his tradition of *l'homme moyen*, and about the one billion people living with misfit conditions, and the odd staying power of ordinary chairs. I wondered if the average doesn't just have to shed its status as an ideal. I wondered whether design might be able to leave behind the average altogether.

In the living room on the day of my visit, Niko's chair looked snug and just right. Mark and Vera were pleased, too, and grateful for this beautifully customized piece of furniture, as a practical object and as a manifestation of interest and investment in their child, down to its smallest features. But Yasmin, the physical therapist, still wasn't sure about the headrest, Vera told me. It wouldn't have occurred to Vera to complain about the chair's smaller, subtler details. It would have seemed ungrateful, she said. But Yasmin knows the body, and she knows Niko, and she knows that ADA can build almost any chair a person can dream up. She wants to make sure the headrest is doing absolutely everything it can.

ROOM.

DeafSpace, a hospital dorm, and
design that anticipates life's hardest
choices. Rethinking "independent
living."

A house that has been experienced is not an inert box. . . .
All really inhabited space bears the essence of the notion
of home. —GASTON BACHELARD, *THE POETICS OF SPACE*

A dormitory at Gallaudet University shows off features based on DeafSpace principles: long sight lines made by half-height walls, solid colors for strong visual contrast, seating for easy in-the-round signing communication.

t was when Maya showed me the benches at Gallaudet University that I started to glimpse sound—the physical structure of it, the elastic bounce of its travel. My friends who are deaf have always told me that sound also belongs to them—that hearing people are forever getting it wrong to imagine deafness as a "silent world"—but the benches were the thing that made this idea vividly real. They were a feature in the design at the scale of rooms at Gallaudet, alongside a dozen other architectural choices that a hearing person could easily miss.

Maya had paused for a moment in our campus tour to point them out, standing in the middle of a big, airy common space lined with windows on three sides, the lobby of a dorm where many students study and socialize, alone or in groups. The benches serve as seating for nearby wood tables, sets that are interspersed with soft fabric chairs arranged 360 degrees around for discussion. "Wood is the best material for this kind of group seating," she told me, and mimed lightly slapping the wood with her palm. The resonance of wood makes it reverberate when struck. Students sometimes tap or slap nearby surfaces to get one another's attention or to call a group to order, she said, and materials like concrete or thick plastics tend to absorb the sound rather than scatter it productively. Nearly all the students on campus, like Maya, are deaf or hard-of-hearing, so the materials and room arrangements make a difference. The benches are where sound does its tactile work—not so much as noise, though it is that, too, but as vibration. Sound as a medium, as a force to harness and amplify or sublimate and send to the background—these properties and possibilities are what I understood tacitly about acoustics but had to see all over Gallaudet in person to really understand. It was one of many ways that I witnessed how deafness produces a distinct sensory ecology.

At Gallaudet, most communication between people on campus is in American Sign Language, or ASL. The campus was founded for deaf education in 1864 and is the only bilingual university of its kind, using both English and ASL in its curriculum but self-described as "a signing community" and a proud home for deaf culture. I had panicked briefly in the taxi there, remembering suddenly that I might have to ask for directions on campus without rudely assuming spoken English, and started to dredge up the ASL alphabet that was still in my manual memory from childhood. I also cued up a Notes app on my phone, for typing exchanges if necessary. But in the end, I found my way to my meeting place, right on time, and to Maya, my tour guide on a brilliant October Friday, who was from nearby Maryland and majoring in interpretation.

Even in jeans and a sweatshirt, Maya had the shy formality of a student on the job, tasked with reciting the facts and figures for an earnestly curious stranger. Maya and I formed a triad with our ASL interpreter, a bilingual graduate student who carefully chose her own positioning in space to be out of my visual field but easy for Maya to watch while also making eye contact with me.

Maya was patient with my dozens of questions, and, watching our interpreter and recalling protocols established with my deaf friends, I remembered to tap her lightly on the shoulder when I wanted to ask her something—a form of touch I might consider presumptuous with a stranger in American hearing culture. The three of us did our round-robin of communications while we traipsed about, indoors and out, looking at old-school plaques and statues along with technologically sophisticated classrooms.

Education for deaf and hard-of-hearing students has been around for a long time, of course, but there's a new, or perhaps new-and-old, or perhaps *renewed* set of ideas about architecture happening at Gallaudet, which I had been reading about for years and finally went to see for myself. For students taking classes, holding club meetings, working and living there, the architecture—walls and furniture and doorways—is designed to support the deeply embodied, three-dimensional, complex social structures of sign. There are the wood benches—an old and commonplace technology, long used as a tool for somatic communication—but, all over campus, in wall heights and atrium structures and color choices, there was evidence of newer designs inspired by deafness: it's called DeafSpace. Gallaudet's architecture emphasizes the particular capacities and *assets* of deafness, an utter reversal of even the most generous and well-meaning interpretation of "inclusion." DeafSpace isn't a plea to "make room" for deafness. It's an unapologetic and joyous expression of the integrity and beauty of deaf experience, codified in a series of strong principles that inform the way the rooms here look and operate and feel.

LLRH6, the Living and Learning Residence Hall where I encountered the wood benches, is a DeafSpace dorm built in 2010. It's a glossy modern building with glass, wood, and metal surfaces and concrete floors, plotted on Gallaudet's Kendall Green amid the ornate decorum of smaller Victorian structures from the original campus and the blocky, pragmatic towers of other dorms built in more recent decades. It's an architectural mix that would feel familiar to families on college campus tours everywhere, and it's possible to miss the subtleties if you're a newcomer.

Two sets of extra-wide automatic sliding doors at both entrances to the building make it easier for two or more people in manual conversation to look not at the door but at each other, entering side by side and signing without interruption. Inside, a uniformed security guard sits at an information desk framed by a bric-a-brac of colorful flyers, as on every college campus, a collage of papers announcing the names and images of the resident assistants, the meetings coming up, the safety reminders. At a coffee kiosk on the ground floor with a blinking OPEN sign in retro neon, students were sitting and chatting, with music from a boom box playing nearby.

The rest of the space is a sprawling, high-ceilinged lobby lit by sunshine through its enormous glass walls. Maya told me that a seamless transition to a naturally lit interior is ideal for visual language. "If you have to adjust your eyes suddenly to a dark interior with low lighting," she said, "it's a big interruption when you're midsentence in sign." We moved through the space, which is set up for small-group meetings or events for large crowds by its division of functions: there's a long graduated ramp that runs the length of the room, and three pods with tables and chairs breaking off from that ramp, all in a row. Each pod is partially surrounded by half-height walls (sometimes called "pony walls" by architects). It's an intimate but open structure, offering sight lines that run the whole length of the lobby, making the visual communication of sign easy at short, medium, and long range. There's no urgent need to block noise here, so walls aren't used as sound barriers.

The half-wall surfaces of the space were painted in a medium sky blue, a color choice that provides maximum clear contrast with a variety of skin tones, so that the finer motions of hands and fingers can be easily seen. I spied solid greens and blues, in paint and fabric, in lots of places on campus, a better choice for communication than busy wallpaper or glaring bright white would have been.

We surveyed the lobby as a whole, and saw how its intimate pods can also easily function as a larger, shared single structure. The gentle cascade of ramp and pods, each level slightly lower than the last, ends at the ground level in a cluster of tables and chairs, with multiple oversize flatscreen TVs hung high above. Arranged this way, with the simple shift of some seating, the room can turn into an amphitheater for large groups. A live guest speaker or a presentation on the television screens, in sign or captioned, is visible from each pod all the way to the top of the room and at the coffee counter. Where the visual language of sign is paramount, these structures and choices are ideal.

When we walked outside the building and paused at the exterior, Maya turned my attention upward. We squinted at the glass walls that partially enclose the dorm lounges on the corners of the higher floors. "See up there?" she said. "If I'm hanging out in that space, watching TV or whatever, and I see a friend outside on the sidewalk, I can easily communicate because the glass makes it possible." She and her friends use their mobile phones for texting too, of course, but the deliberate choice to remove a visual barrier—what might otherwise look like a standard modern choice of a glass exterior—takes its shape from the affordances, not the limitations, of sign.

DeafSpace is a series of more than a hundred design elements that were the result of collaborative research conducted by Gallaudet students, faculty, administrators, and campus architect Hansel Bauman starting in the mid-2000s. Bauman—who is bilingual and describes himself not as DeafSpace's inventor but as its "finder" and its "keeper"— worked with this assembled research team to bring close attention to the ways deaf people have always used architecture, their innate and implicit knowledge. The group canvassed the entire campus, looking for clues old and new about how deaf people already use architecture in an adaptive way. They organized their findings by a series of categories: (1) sensory reach, (2) space and proximity, (3) mobility and proximity, (4) light and color, and (5) acoustics.

As carried out between speakers, ASL is deeply spatial, playing out not only in the body but in an extended circumference around the body. Words and phrases take on complex connotations and even radically different meanings depending on the energy and dynamism of arms and hands and facial gestures and the rest of the body—degrees of emphasis and context that can make the difference between earnest expression and sarcasm, a whisper or a loud demand.

The spatial implications grow with the size of the group conversing. Everyone needs to be able to see everyone else who's speaking, including those who might pipe up from the back or elsewhere in the wider visual field. There's a bigger spatial dimension to a deaf person's entire communication apparatus. Imagine a 360-degree circle around any set of people in a conversation when using sign: an invisible bubble of space where language is both expressed and received, and one that's bigger and wider than for those using spoken language. This full circle of communication is part of the "space and proximity" principle in DeafSpace—guiding its recommendation for clear or frosted glass on doors, for example, so that speakers in a hallway can see a third person approaching from behind them.

Bauman, with a sweep of silver hair, worker's khakis, and a wooden beaded necklace, walked me around the campus, demonstrating the tacit spatial agreement that would have to be in place between us if we were signing instead of speaking. As we walked a length of sidewalk with cars parked perpendicular to its edge, he turned to me, his pace still brisk and his eyes trained on mine as they would be for signing, and said: "Now—what's about to happen here?" I'd spotted the hood of a car in my peripheral vision, parked so close to the curb that its nose jutted far into the sidewalk. "I need to alert you that this obstruction is coming up," I offered. "Right," he said. "You need to physically reach out and cue me that it's in my path. I'd be pissed off if you didn't." That social awareness—not just for oneself but for the other—is a naturalized set of behaviors that come with the social work of sign.

In a group, one person will often take the role of wayfinding, scanning the environment for barriers or approaching changes in direction. The designated guide will interrupt the flow of conversation among the group only when necessary. Design for the DeafSpace category of "mobility and proximity" can do its work here, too. DeafSpace recommendations include extra-wide hallways that allow for sustained conversations without group collisions and stair treads with extra depth that make for a slower ascent or descent, built to accommodate ongoing talk at a more leisurely pace without tripping. Ramps in place of stairs, where possible, also mitigate the focused stepping that detracts from signing in motion; ramps also extend access, of course, to people who use wheelchairs. Matthew Malzkuhn, a Gallaudet alum and researcher on deafness and design, has said that this three-dimensionality in language and the body is integral to what it means to be deaf. "[We're] spherical people," he told an interviewer in 2007, so it makes sense that the architecture here should beautifully bear out deafness as a way of being.

These skillful accommodations to the visuality of sign language are out of sync with many standardized features of rooms and buildings. Think about the ones in your own classroom experience: tables for meeting that are squared off at the ends, creating some midlength sight lines but blocking the people seated at the corners. Many classrooms still have chairs or desks arranged in rows, presuming a simple two-way sight line between a teacher in the front of the class and students facing her. Rooms are often set up too much like tennis courts, with their back-and-forth exchange, and too little like a card game, where everyone's eyes watch the faces and hands of everyone else. At Gallaudet, classrooms are always set up in a U shape for that card game effect, a relatively easy adaptation. But there are other structures where the contrast is more pronounced. Hallways in many typical buildings are built at right angles, presuming the audible cue of footsteps to alert a hearing person to someone approaching from the opposite way. New hallways at Gallaudet are built to curve gracefully around a corner, extending the line of sight and avoiding collisions. That's part of its principle of *sensory reach*.

Like Malzkuhn, Bauman describes the "spatial kinesthetic" of sign language as particularly conducive to architectural thinking. Signing is three-dimensional by nature, so it implies not just words in a line, but heights and widths and depths for making language the rich repository of communication that it is. The unidirectionality, even "linearity," of spoken language seems paltry by comparison. Never had my spoken English felt more comparatively impoverished than on my visit. I was painfully conscious of fumbling for what suddenly struck me as the pallid choices of a word-based vocabulary, and of how my every wispy utterance seemed to disappear immediately into thin air as soon as the words exited my mouth. I was aware then of what Bauman's brother, Dirksen Bauman, head of the Deaf Studies department at Gallaudet, has called "deaf gain," a pointed corrective to the commonplace usage of "hearing loss."

It might surprise a hearing outsider to encounter "Acoustics" as a category in DeafSpace, but the qualities of sound are also carefully considered here. For users of cochlear implants especially, infrastructural noise like air-conditioning and ambient noise in clattery, echoing spaces can be very disruptive. At the same time, sound is also of practical use, as with the wood benches, and a big source of enjoyment, too. On and off throughout the day of my visit, the coffee counter in the residence hall played loud music on its boombox—sometimes *very* loud music, turned way up to broadcast its vibration, especially the low-frequency pleasures of rhythmic bass. The pleasure, of course, is determined by the day and the hour and your auditory and somatic makeup. It was too loud, at times, for me to hear my interpreters, but a beautiful bit of sensory input for folks who don't primarily use their ears to understand one another. Loud music is such a daily feature of life here that some residences are designated as "quiet dorms," with set curfews that limit the hours music can be played, out of respect for fellow students trying to work or sleep.

I already knew that my deaf friend Mel relishes her private concerts when driving solo on the highway, the volume turned up as loud as it will go and her left knee held against the speaker at her side to absorb its vibration directly, the better to joyfully howl along with the Beatles or Stevie Wonder songs she's memorized. After the Gallaudet tour, I could better imagine what that might be like. Sound is a multisensory pleasure because it's a wave. That's why it's something you can feel—the diffusion of its energy from a high concentration to a lower one as it moves through space. Whether from that hand striking the bench or the tiny fluttering motions of paper created in the woofer of the boom box, it was the waves that became apparent and palpable to me, running invisibly through all the atomic weirdness of every structure in the room: the wood, the glass, the lining of human viscera in a delicious thrum. It's like when you drag your finger through a puddle of water, injecting the surface with a sudden energy that spreads a pattern of ripples out in a fan, visibly cascading from its source to the world.

For many at the time of Gallaudet's founding, deafness would not have been described by its assets, because deafness, writes historian Douglas Baynton, "is not what it used to be." *Mute* and *semi-mute* and *deaf and dumb* were the terms commonly used to variously describe conditions of deafness up until the nineteenth century in industrialized cultures. Samuel Johnson called it "the most desperate of human calamities." The historical literature of that time often describes deaf people by the strangeness of their gestures and in pitying terms. But by the nineteenth century, Baynton writes, firsthand accounts of deafness by deaf people themselves offered a different narrative. Harriet Martineau, a deaf British woman, wrote in 1836 that while she had found deafness "almost intolerable" as a young person, she came to understand her early experience as a "false shame" that she shed in adulthood. Baynton cites another British writer, John Burnet, who became deaf at age eight, as anticipating a social model for disability in his *Tales of the Deaf and Dumb* in 1835: "While deafness 'shut its unfortunate subject out of the Society of his fellows,' this is due not to being 'deprived of a single sense,' but rather to the circumstance 'that *others* hear and speak.' Were anyone to use 'a language addressed not to the ear, but to the eye,' [Burnet] maintained, 'the present inferiority of the deaf would entirely vanish.'"

Schools for the deaf were founded in urbanizing, industrializing cultures in the eighteenth and nineteenth centuries. The National Institution for Deaf-Mutes, established in France in 1776, was widely influential in bringing specialized learning to deaf students internationally. But its success collided with resistance in the form of late-nineteenth-century nationalist campaigns to homogenize speech and language. The pressure was strong and effective: a reactionary and ultimately eugenic push for "oralism" swept through modernizing Western cultures, forcing educators to emphasize the use of oral speech for deaf students wherever possible. Signing, as a minority language, was discouraged as a threat to the unifying work of national culture. The Milan Conference of 1880, an international confederation of teachers for the deaf, agreed as a group on "the incontestable superiority of speech over signs." The effect was long-lasting, with oralism as the default mode of mainstream deaf education until the 1970s, writes Baynton.

Deaf culture became recognized *as* culture and a conscious political identity in the United States over the course of the 1970s, with deaf activists insisting on the integrity of their ways of being, without the need for spoken language. Gallaudet was closely tied to these efforts; a renewed interest in sign language had swelled on campus starting in the 1950s under linguist William Stokoe. Activists made a strong case for their language, their history, and their culture as a fully normal variation on human existence; some took on *Deaf* as an identity, capitalizing it as a proper name on a par with ethnicity.

To be deaf and politically active is to reject the long history of "curing" the absence of hearing, whether by insisting on oral speech or by investing only in technologies that augment hearing rather than in the linguistic and cultural norms of the community as it is. Gallaudet was instrumental in this legacy of rights. In 1988, students insisted on ousting a recently elected hearing president and hiring the university's first deaf president. They demanded that a full 50 percent of Gallaudet's board representation be deaf, by policy, and that they be guaranteed protection from punishment for these acts of protest. The march posters from the era express the strategic alignment of deaf activists with other civil rights groups: "Jews, Blacks, Women, and Now It's Time for Deaf" reads a hand-made sign from a rally in Gallaudet's archives, a stop on my tour with Maya. It's a pivotal moment in the university's history and a specific align-ment with the legacy of minority rights in the United States, the sum of which is a philosophical claim and demand: that the condition of deafness in the world is not a personal matter only, but a social and political one; that the deaf are a marginalized identity group who deserve equality.

DeafSpace, then, in marking and drawing up buildings around its culture, is an indirect outgrowth of this half century of political will. I certainly felt that way talking with Bauman, whose conversation spanned from the details of materials and shapes of hallways to the big picture of what it all means. He's an architect's architect; he told me he's interested in how buildings can help do the metaphysical work of what the philosopher Martin Heidegger called true *dwelling*—a social triad made of people, one to another and in their environments. DeafSpace isn't prosthetic ar-chitecture, at least not in the sense of replacement parts. It's architecture that's wrapped *around* the otherwise invisible, taken-for-granted ways that relationships, mediated through the sophisticated, beautiful lan-guages and behaviors of sign, are built. The visual evidence traces the history of deaf politics, housed in the most playful and compelling form of a question: What is deaf design?

Since its 1864 charter by Abraham Lincoln, the campus has always borne some features of design that were built with deafness in mind. In one of the original buildings, there was a vibratory doorbell—a handle on a chain that a visitor would pull and release to drop a lead weight housed in a chamber that would drop to the floor. The impact of the fall and collision would be felt somatically by people inside, registering the visitor's presence without relying on sound. It was a clever workaround built on sensory substitution and mechanical problem solving. There are aspects of the current design that mirror this design decision: visitors to faculty offices announce their presence by flipping a light switch outside the door in lieu of knocking. A textured cobblestone border along a sidewalk alerts walkers in conversation that a curb is nearby. Several buildings, both old and new, make use of atrium structures to maximize sensory reach—extending the possibilities for visual language not just across a room but between floors wherever possible. Maya showed me the wrap-around stairs inside one of the original boys' dormitories, first built for high school students. The students would drop their dirty clothes down the column of empty space at the center of the winding stairwell; when the laundry was clean, each boy would stand on one of the steps, passing the piles of clothing up the chain of floors with a clear line of sight for orchestrating the task in every direction.

These architectural features arise from deep and creative adaptation to a misfit condition. But it took a political push by deaf people—to be recognized not as a medically defined community, taught and managed by hearing people according to ideas about hearing loss, but as an identity—to create DeafSpace, which recognizes *deaf gain* in rooms and furniture and buildings.

Given Gallaudet's self-described identity as a "signing community," it might be surprising to learn that a full third of Gallaudet students arrive each year as transfer students, and between 15 and 20 percent are new signers. There's a boot camp at the beginning of the year, four weeks of intensive training in ASL for any student who needs it, and four levels of ASL classes beyond that. The university provides translation services for students for up to a year of enrollment. Students come to the school from all over the world, but even those from the United States arrive with a wide spectrum of familiarity with signing. During my visit, more than one student described to me the "culture shock" that greeted them, as their brains struggled to ramp up to a speed that would allow them to communicate with others. Some students are longtime signers with other deaf members of their families; many others were raised to do some combination of lip reading, voicing, or speaking SEE—Signing Exact English—a sign language with a grammatical and structural shape completely unlike ASL. As with spoken language, there are hundreds of sign languages spoken globally, and they vary widely. British Sign Language, for example, bears little resemblance to ASL.

One student told me that she learned ASL only after spending a year on campus working hard at it, and that her hearing parents and family still don't sign at all when she returns home. Even attending Gallaudet was a contested matter with her parents, she said, but a full scholarship gave her the autonomy to make her choice.

Another student, an American and native ASL speaker with a number of deaf family members, described the necessity of *code switching*, not just with students who aren't experienced signers but with faculty and staff as well, some of whom sign fluently and others very little. The code is partly linguistic and partly political, she implied. Some students fluent in ASL tolerate the need to switch with more grace than others. Some professors emphasize written academic English in papers, and others lean more heavily on using ASL for reportage. (Students use the computer labs to do desktop work but also to make short films of themselves signing for their reports, to upload and send on to their professors.) All of these adaptations to communication in these various forms are, in a sense, code switching that grapples with a larger question: How much should students work for mastery in the language and mores of the dominant hearing culture, and how much should they cultivate their deaf—or Deaf—identity?

The architectural design here, new DeafSpace structures alongside the old buildings, created a visual landscape that I came to see as reflective of the tensions in the interior work students are doing: not just living away from home for the first time but living with their linguistic and social norms abruptly reversed. They're part of a generation for whom inclusion, via mainstream placements in public schools, is much more the norm. Disability advocates generally insist on this form of inclusion, but at Gallaudet, I saw the trade-off. Do these students come to campus to in some way *become* deaf? I found myself wondering, after conversations with students, whether at least some arrive on campus to newly foreground this piece of who they are. Their choices all take place in this institution that fought so hard for the education of deaf people in the first place, and an institution that's had to fight for its students' opportunity to learn ASL while the politics of identity wax and wane. Perhaps DeafSpace, then, is an example of what architects often call a building: an "envelope," but of a particular kind—a flexible enclosure for human action and interaction meant to mark and trace what's hiding in plain sight. Imagine the envelope being drawn up from the ground, sides and corners and rooftop to form a shape *around* a series of habits and patterns for living that deaf people have been doing forever—a housing for the extant subtleties of these ways of being.

When I talk to a certain kind of know-it-all hearing architect about DeafSpace, I get one of two common responses: either they rush to insist that the broadly open-plan features, circular seating structures, and emphasis on natural light are "just good design" and have little to do with deafness after all, or they want to poke at the premise, suggesting that DeafSpace must automatically be working at cross purposes with other kinds of accessible design. If it works for deaf people, will blind folks be left out? (The short answer is no—Gallaudet also educates "deaf plus" students who are blind, or use wheelchairs, or have other kinds of complex bodily states.) But I think they're missing the point. The buildings here do a very good job at matching the functionality of the dominant culture, yet the idea of DeafSpace isn't so much about the features—what the design looks like—but instead about the qualities of attention that DeafSpace practices as its *process* of design. The granular, intimate details that make up the patterns of culture and communication happening outside the normative human sensorium are an invitation: to look closely at what communities people build, outside the mainstream, and to formalize an often invisible but long-held tacit knowledge of the world. DeafSpace, on display at Gallaudet, offers a way for architects to keep returning to all the languages of the body, to how we all might hear and see one another better; to prioritize relationships and the social dimensionality of embodied languages; to reexamine the standard structures for living and ask where opportunities for refashioning might make a more livable series of rooms. Surely the rooms built on DeafSpace principles embody what philosopher Gaston Bachelard means when he writes that "all really *inhabited* space bears the essence of the notion of home." To inhabit—from its roots in *habitare*, or "to dwell," which has its further roots in *habere*—"to hold or have," or even "to give or receive." The giving and receiving that is so acute in spherical, socially interdependent sign makes a habitable home for the students who gather there.

At Maya's urging, I visited a site off campus, one that might be the most revealing work of DeafSpace: the Signing Starbucks, on nearby H Street, a short walk from campus. Opened in 2018, this Starbucks appeared to be like every other Starbucks, but for the lightest touches that reverse the norms of the room. When I approached the counter, I didn't need elaborate instructions; I was able to infer from the equipment and others in line before me what to do. The staff signed to one another and greeted me with the polite "Hello" of a short wave, palm against the temple. I wrote my order on one of several tablets with an easy wipe-away surface, plus my name. I paid by card in the same way I always do, lightly gesturing with the cashier in thanks and payment directives. When another staff member interrupted my cashier in the middle of my order, I took in the way she started her sentence with a polite acknowledgment of me—a sign that registered "Excuse me" by following the gesture of my writing to indicate who she was speaking to, and then proceeding to sign with her coworker. I understood, in context, both her politeness to a customer and the ordinary coaching that a new hire needs. When my order was ready, a screen mounted at the end of the coffee bar announced my name in a list.

The Signing Starbucks works by implementing the subtlest possible changes to the technologies of the room—not even particularly high-tech gadgets are needed—and in an instant, all the power dynamics are quickly inverted. It's what in my field we'd call *service* design: a combination of elements that includes products and interactions to make a desirable experience, in this case a customer transaction that's intuitive and frictionless. I needed no extra instructions to be a customer, so you might say the place was nearly universally designed, and yet the storefront operates entirely in a spatial language that relatively few people know and understand. It's a signing community in the middle of the hearing public square, visual evidence that spaces can easily be shifted, expanded, and above all *shared* in unusual and multidirectional ways.

It was clear on my visit that the store is a source of pride on campus; there are plans in the works to open a bank branch and a pizza parlor, each with a similar service structure, in the neighborhood nearby. I took my tea to go, through a customer crowd who were roughly half voicing and half signing, and out into the street again. I thought once more about the real unresolved provocation of the misfit condition—who becomes a misfit, and where, and how, can change in an instant or at a glacial pace.

And how telling to find it in both places: in a room selling the world's most mainstream commercial coffee, and on a college campus where the curriculum is all about becoming.

Making big choices about who you are and which worlds you belong to is the story of many college students: it's the foundation-shaking work of late adolescence and early adulthood. But the most interesting creativity often results when there's an unusual and urgent degree of friction in the meeting of body and world—whether that friction is born of capacity or history or demographic background. In disability history, the dorm room has sometimes been the designed environment that launches the biggest questions, personal and political, about how we decide to live.

In 1962, Ed Roberts, a polio survivor and wheelchair user whose body required a breathing apparatus by day and the shelter of an iron lung at night—an experience he loved, like "resting on a cloud," he said—applied for admission to the University of California at Berkeley. His determination to find a way to live on campus became an unusual story of adaptive architecture that would spawn an entire international movement for independent living and dwelling and a way to redefine what it means to ask for help.

Roberts was accustomed to workarounds for getting an education, having attended high school from home in the 1950s with the use of a phone extension connected directly into his classrooms. His mother had insisted, despite protests, on Roberts receiving a mainstream education. (Even so, he and his mother had to petition for his diploma at graduation; it was initially held back by his lack of driver's education training.) After a stint at community college, and with the encouragement of professors, Roberts applied to Berkeley and was admitted at age twenty-three. There was no box on the form to check that would indicate he was disabled, so, after his acceptance, he had to figure out how he would live on campus.

WHAT CAN A BODY DO?

Students with other disabilities had attended Berkeley and other colleges in the United States, but nearly all of Roberts's body was paralyzed, so he would need assistance in an acute way, for things like getting from his bed to his wheelchair every day—the kind of help that exceeded dormitory norms.

Berkeley administrators initially tried to backtrack and refuse him admission, citing past failed experiments with enrolling disabled students. Roberts went to see Dr. Henry Bruyn at the Cowell Hospital on campus. Bruyn had overseen the care of children in Roberts's generation who'd contracted polio and lived with the aftermath of the disease. He agreed to work with Roberts to outfit a hospital room for his needs while on campus—as a makeshift, semipermanent approximation of a dorm room. Berkeley agreed, and Roberts arrived on campus, but expectations were low. The headline in Cal's own newspaper read: "Helpless Cripple Attends UC Classes Here in Wheelchair."

A hospital as dormitory: a new idea that required deep imagination. Outfitting the room for Roberts's medical needs was easy. The crucial matter was how he would do what all college students have to manage if they move away from their family home: manage an open schedule, make decisions about balancing school and social events, forge new relationships, take care of their bodies. For Roberts, that would mean creating and managing relationships *about* the care of his body—hiring and coaching and firing personal care attendants to assist with intimate tasks like dressing and toileting, making weekly schedules for transportation, and more. These were brand-new skills for Roberts. But the arrangement worked, and by 1969, a dozen students lived in the Cowell Hospital dorm in similar circumstances, each of them a wheelchair user orchestrating a new life as a student and as a household caregiving manager-for-one.

Historian Bess Williamson, in her book *Accessible America*, cites oral histories from the era that evoke the high stakes for these students. Herb Willsmore, who enrolled in 1969 and lived at Cowell, said that the dormitory rooms—and the newly independent lives they set in motion—were the catalyst for much bigger conversations among student residents there:

> We started talking about our rehab experiences, and the fact that they didn't teach us how to transfer in and out of a car, or how we were going to manage once we got home, and how to set up a room or a house to where it's workable for someone with a disability like ours, or how we were to empty our leg bags, or what it's all about, hiring attendants, how you interview, all those kinds of things.

Another resident, John Hessler, wrote that by the time he completed his stint at Berkeley, he had been "an employer for six years of orderlies, attendants, cooks, maids, secretaries, and research assistants." It had been an education that far exceeded the walls of classrooms and semester structures. The dormitory began to stand for something else symbolically, something more than getting a college degree. The dormitory bond became a political identity on campus: a group of disabled students evolved into self-advocates, insisting on more than admission. It was the 1960s in Berkeley, California, and these students understood that their existence on campus was closely tied up with civil rights—theirs alongside other marginalized groups newly awake to political strategy.

The dormitory group called themselves the Rolling Quads, a slangy shorthand for *quadriplegic*, and, fortified by their numbers in the hospital dorm, made their presence known with visible protests and gatherings, generating demands for disabled rights on campus by convening political meetups for "crips and walkies"—a mixture of students who used wheelchairs, students who were blind, and their friends and allies who used legs and could see. They created a formal lobbying group on campus, the Physically Disabled Students Program (PDSP), and got needed services funded first by a federal grant and later by a clever budgetary design: they proposed and won a twenty-five-cent increase to student tuition at Berkeley to cover referrals for housing and attendant employment; wheelchair repair services on campus, including mechanics who were knowledgeable about the new-to-market motorized wheelchairs; and more.

Other disabled students made the connection: Must we accept and bend to the norms of the institution? Or can we insist, together, that the institution also bend for us? The PDSP provided services that were more than practical. They were articulating the seeds of a new definition of independence entirely, one that transcended wheelchair use. These students were taking stock of the built environment—classrooms and streets and homes—and looking for ways those structures could be made more accessible. At the same time, they rejected the rehabilitation paradigm in which they'd grown up—in which their independence was measured only by their capacity to do things literally for themselves. They maintained that the need for physical care would always be present in their lives, but they insisted on a way to live with chosen forms of help while keeping decision-making power. Instead of defining independence as "self-sufficiency," the standard for independence in the clinical settings where they'd been treated as patients, they claimed that their independence would be understood instead as "self-determination." The difference separated the dignity of authority and choice from the action itself. Asking for and receiving help with self-care tasks like buttoning a shirt, for example, was understood as a high degree of *dependence* in a rehabilitation paradigm. But if a person needed fifteen minutes of assistance with the shirt and with getting out the door to the bus, that person would be less dependent than a person who took two hours to dress on their own and could not leave the house. Uncoupling *assistance* from *dependence*—or perhaps bundling assistance together with a richer idea of *independence*—changed everything for these activists, because now they could press for a whole array of products and services that would support a desirable life. Judith Heumann, one of the instrumental voices for independent living, said in 1978 that "to us, independence does not mean doing things physically alone. It means being able to make independent decisions. It is a mind process not contingent on a normal body."

Roberts and the Rolling Quads were so much in my mind when the world slowed to a stop under threat of COVID-19 in early 2020. My own campus and my children's schools abruptly shut down and went to a distance learning model; we all adapted, shakily, in fits and starts, to the kinds of remote connections that Roberts had known as his "normal" for years on end, all through high school and community college. What powerful, prescient expertise was right there in my country's own history. On the day before we all scattered, I told my ninety design students about Roberts, about Cowell Hospital and the powerful redesign that was now their inheritance for learning at a distance. I didn't tell them to *be grateful* in light of the struggles of the Rolling Quads, to disregard their own losses that would come with the shutdown of our physical campus. On the contrary—I told them to see the connections between those examples half a century ago and the adaptations that they were now taking up: an invitation to resilient new forms of designed interaction. There would be closures and openings—that we knew for sure. But could we stay awake to the work of adaptation even as it unfolded?

The PDSP, for its part, eventually exited the boundaries of the university campus and opened a storefront in downtown Berkeley. There, in 1972, in the square footage of a closet space, a new organization emerged: the first Center for Independent Living. It operated as a nonprofit that provided the kinds of services for the public that had first been prototyped by the student group: referrals for housing, attendant services, and employment help. The brick-and-mortar organization, replicated in other cities in the United States and eventually all over the world, became the Independent Living Movement (ILM). Based on that early advocacy work, federally funded Centers for Independent Living now exist in all fifty states and have been replicated internationally; they continue to provide these services, on a drop-in basis, for people who need them. And today, in a building named for Roberts at Berkeley, every visitor is greeted by a joyous, monumental, bright red circular ramp that winds from the bottom floor up to the second. I spent a scant half hour in the presence of that ramp one late afternoon and saw a small parade of bodies all using its physics for their travel on campus: a commuter with a bike, a parent with a small child, a person using a motorized wheelchair, and plenty of walking folks on two legs.

For the last couple of years, I've been visiting Steve Saling in Chelsea, Massachusetts, who helped me see the long inheritance of the ILM in the home he designed for himself and two dozen others, a mix of technology and human care that he created to arrive in time for his body's big changes. Steve got a diagnosis of amyotrophic lateral sclerosis (ALS) in his early forties. Now, more than a decade later, Steve lives with the ILM's expanded idea of independence. But he also lives with something we all share: not just independence but the plain fact of *dependence*, too. In Steve's life, I saw a constellation of designed gear and services that changed my perspective and my vocabulary about assistance—about human needfulness and its role in a desirable life.

From the lobby at the Leonard Florence Center for Living, there's a set of extra-large elevators, the kind a full-size motorized chair can enter and exit, for taking visitors up to the residence floors. On the days Steve met me downstairs, he directed his chair's movements by the padded switches that extend from the headrest to frame his temples; he moved his head very slightly to the left or right to make the commands. These motions, along with some muscle movement in his face, are the only mobility he has now. Steve always greeted me with a combination of his smile, his blinks and light nods, and his words—"It's good to see you, Sara"—expressed in the voice of an automated text-to-speech device on his chair. He then summoned the elevator himself, taking us up to the third floor, where he lives in a series of rooms that look and feel like a home.

There's a cheery yellow façade that frames the entrance at Saling House—an interior wall painted to look like the exterior of a suburban house, with real wood shutters around a "window" next to the entrance, a couple of potted plants, and a welcome sign that bears the house's name. Two sets of oversize doors open to the interior, each pair sending one door swinging in and one swinging out for the widest possible berth. Steve and I crossed into a shared common area just inside the main entrance, where the furniture is like that of many living rooms in single-family homes. The chairs and furniture, the colors of the walls—all of these are choices that have been made to mimic the hues of nature: warm yellows and soft blues and greens. The decor, including a fireplace, framed artwork, and club chairs, follows the logic of what lots of people instinctively do when they furnish their own spaces: choosing lamps and furnishings whose simple curves mirror those of the body, the visual language that says *Come and relax*, not *Beware the machines*. The space is so homelike that you might miss the extraordinary sophistication of the technology that's everywhere.

Now fifty-one, Steve is eleven years into his diagnosis. He has the boyish affect of someone much younger and the chin-length brown hair that he's had all of his adult life. He has the expressive eyes of someone whose other facial muscles are now slowed almost to a stop; the eyes do much of the emotional work for him, and they're good at it. Steve operated the elevator, opened all the doors for me, sent texts to other residents, and lowered and raised the blinds, all powered by wheelchair commands. All of the heating and air-conditioning, the music and media equipment, and much else at Saling House is controllable by software that's available, to Steve and to the other residents, via screen interfaces. All the residents of

Saling House have either ALS or multiple sclerosis and need adaptive equipment; most use wheelchairs. On each chair, there's a tablet that functions as central command, manipulable by hands or soft-ended sticks moved by their mouths, or, in Steve's case, by the tiniest cursor you've likely ever seen. On Steve's eyeglasses, rounded classic gold frames, the cursor sits right in the center at the bridge of his nose. It's like a flat round stud earring, easy to miss. His head movements direct its connection to the tablet in front of him, and that's how he drives all the infrastructure. It's the master key to this residence made for bodies like his.

On our visits, Steve and I would settle into the living room, or at the long table where meals are shared, or in his suite, where his walls, painted in jewel tones, are covered with images of himself in years past, hiking and climbing, and in the present, doing adaptive scuba diving with a team around him for spotting and tech support. I would ask him questions, notebook in hand, about his daily life, about the history of Saling House as an architecture project, about his plans for the future. He would type the answers, using the cursor. On that tablet screen, imagine a floating vertical line of the alphabet rising from the bottom of the screen's edge, repeating itself over and over as it travels up the monitor. It's a river of letters, endlessly repeated, hovering and sliding in a liquid manner, like cells under a microscope. Steve's cursor motion "catches" a desired letter as it floats up to the center, separating it from the fast-moving line. A text prediction algorithm jumps in to guess the likely word he's going for, making shortcuts and lacing together sentences based on common phrases and his history of communications. But to watch the whole thing is like a trippy crossword puzzle for the uninitiated; it's taken a studied integration of body and machine for Steve to make it work.

At certain points in our back-and-forth, Steve would hit play and the computer-automated voice both spoke and displayed the words to me on the reverse side of his monitor. And in the interim, business as usual would continue: a building maintenance worker interrupted us to check with

Steve about a decision on the water heater; he blinked his assent and returned to typing, pausing to greet nurses or other residents when they passed. This is Steve's daily life in his altered body, and he's careful to assure the world that some things haven't changed. There's a button mounted on the back of his headrest that instructs strangers to speak to him as they would anyone else. In the center it reads *ALS has stolen my body, not my mind*. And framing the edges are the words *I am living with ALS*. It's a fact and a declarative statement—not *coping* with, not *managing*, not even *suffering from*, but *living with* this condition and every new thing it brings.

Steve was a practicing landscape architect when he started experiencing some neuropathy and a loss of motor control that got him worried. After a series of doctor visits, he got an official diagnosis in late 2006. He paid particular attention to medicine in those early days, as would anyone in his situation, but nothing worked. "I tried lithium ten years ago," he told me, "but I don't like the emotional roller coaster of getting my hopes up." Every couple of years, he said, there's the new promise of a possible cure. He mentioned one drug that's said to slow down the onset of the disease, but it costs $90,000 for a year's supply. He's stopped paying as much attention to it. "When something's proven to make a difference," he said, "I'm sure I'll hear about it."

In the early days, while part of his attention was devoted to medical research and pharmacological claims for cures, he also set about doing the work he knew well—the work of design. He was able to mentally fast-forward and consider the harder questions; he started planning a place where he could live when he could no longer stand, or speak, or eat unassisted. Steve took it upon himself to research and imagine the look and operations of a space that could anticipate his future body. If he would lose nearly all movement, he told himself, then he would look to design—his background and training—to make a desirable life possible. Steve partnered with Barry Berman, a leader in nursing home care. As part of the Green House Project, a network of clinicians and caregiving executives who are creating (or restoring) high-quality environments and services for older adults, Berman understood immediately the importance of the idea. The two worked together to find the philanthropic funding, the architects, and the software engineers to bring Saling's vision to life. Those years of planning were difficult; "smart" home appliances had yet to come to the mainstream market. But Saling House opened its doors in 2010; in 2016, a second residence followed.

It's not just the software that's remarkable. The architectural choices are ingenious, too. The kitchen is open-style, with the stovetop and surfaces at wheelchair height, with a big island attached, as in a home environment, but with careful accessible choices like stovetop knobs mounted on the side of the cooktop, rather than on the surface. There are two large dining tables nearby, one just a couple of inches higher than the other to accommodate different styles of wheelchairs. And just off the living room, connected by a screen door, is a patio and garden area, a small open-air oasis up on this third floor of the building. The design features of this landscape may be the most ingenious of all: where the patio ends, there's a patch of green lawn, bordered by flowering plants, but the lawn isn't just grass. It's paved underneath with an invisible bed of recycled plastic, making it strong enough to hold up under the grippy traction of motorized chair tires without getting stuck or tearing it to shreds. It's fitting that Steve thought of this detail, as a landscape architect. Even the sod has been considered with wheelchair use in mind.

Most people with advanced ALS live their lives in chambers that are the anonymous and clinical habitation of hospital rooms: antiseptic, devoid of any homelike atmosphere, organized solely around medical management. Steve didn't want that future, but a dozen years prior to our meeting, no alternative existed. He pursued a prototype that would make for a desirable life in his changing body, and this is what I spent time with him trying to understand. If you got a diagnosis for which there's no cure, and you knew it would take all of your motor control but not your mind, or your ambitions, or your interests—how would you design for a new body, an evolving new life that you never expected, a life that's still to come, out in front of you? Where would design be useful, when no cure and no easy replacement parts are available? How would you want to wake up every day? Steve is integrated with his machines in an elegant way, if people can see it as such, and to watch him in action is to marvel at the environment that he's constructed for himself: a livable residence, animated by machines. "Until medicine proves otherwise, technology is the answer," he's said repeatedly in interviews. It's a slogan for the whole house, hung on a handmade sign near the front door. Call it anticipatory design.

And call it, yes, *independence* too, in the Independent Living sense, mediated by technology. But the digital tools for self-determination weren't the only striking things about Saling House. Lots of ordinary materials assist life here too. There are pads on the footrests made of kitchen towels or soft foam—bits of fabric tucked between flesh and metal parts, to cushion the skin of legs or feet. Steve's brown Birkenstock sandals had strips of sheepskin between the straps, placed just so to prevent chafing. There are soft fabric bags with multiple different-size pockets that can be easily looped with Velcro to the side frame of a wheelchair. Most residents had one or more—a kind of customized luggage, lightweight and multipurpose for carrying supplies. There are the beeps and chirps of medical electronics throughout the residence—gear that's used for monitoring oxygen or heart rate, as you might expect—but much of that gear is housed in curtains of patterned fabric, to keep the clinical coldness of their design from taking over the house. The more times I showed up there, the more I also took note of these subtler adaptations that were conceived by residents themselves, with the therapists and staff who work there, or with family members or friends. These low-tech adaptations were the evidence of deep attention and human care, to an advanced degree that's as intimate as care can be. A mixed ecology of life with machines and care is what makes the independence here work.

These residents face some key choices about when their needs for assistance, mediated by either technology or caregiving, exceed their ideas about the kinds of autonomy they need to feel that their lives are worth living. I asked Steve about the relationships with caregivers there. "Most people, myself included, want to do for themselves as long as they can," Steve said on one visit. "Sometimes, doing for ourselves becomes dangerous for ourselves and even dangerous for our caregivers, so it's important to be able to do an honest self-evaluation and be willing to compromise." Steve told me stories that made it clear that self-determination is taken seriously at Saling House, even when there are dire consequences that would strenuously test the ethics meter for many of us. "Even when dangerous, the staff here are good about giving us residents the freedom to make decisions for ourselves if it doesn't put others in danger," he told me in an email. "A good example is a guy who lived here who was warned about his ability to swallow. He was still able to feed himself and choked to death at the dining room table, eating a bagel. It was sad but he died on his own terms doing something he loved."

But others, like Steve, accept the assistance of human help, and then opt in or opt out of two kinds of machines that make their dependence more pronounced and consequential: Once he could no longer chew and swallow, Steve opted into using a feeding tube that gives him the nutrients he needs without the act of eating. And eventually, he will have to decide whether to use a ventilator to outsource the act of breathing. Some residents choose to depend on these machines, and some don't. These decisions strain the edges of even the most expansive ideas of independence. Perhaps that's why ALS in a meaningful life is so underimagined: most conversations about the condition center on the means of a compassionate death, the ethics of assisted suicide, the autonomy of choices for the end. Those are important debates, too, but Steve taught me to see the life available in ALS. With the ILM in my mind's eye, I saw these choices for Steve as variations on independence, evidence of self-determination. But I saw in him, too, that dependence—needfulness—is also the state of the human.

It is a fact that humans all have a period of extended dependency at the beginning of life and during recurrent periods, such as when they are injured or ill or too frail to fend for themselves," writes philosopher Eva Feder Kittay. "Some prefer to speak of interdependence," she continues, but "we cannot acknowledge our interdependency without first recognizing our dependency . . . in some ways we are simply dependent and unable to respond to the other's needs." Kittay sees the ILM as an important early wave of disability activism, reclaiming the rooms of the home as one part of a newly autonomous way of life for people like the Rolling Quads, creating a crucial redefinition that rejected an *imposed* dependency in the lives of disabled people.

But Kittay writes that independence as an ideal isn't enough on its own. "It is part of our species typicality to be vulnerable to disability, to have periods of dependency, and to be responsible to care for dependent individuals." And indeed, anyone who has thought much about aging, for example, is sooner or later forced to grapple with the question of independent living, including questions about residential design. As we age, and if we have choices, will we want a house in a familiar community or a more fully serviced retirement center? A set of steps outside as a trade-off for having a dwelling all on one floor, or an apartment building with an elevator? A room in a house of your own, perhaps with a dog for security, or a shared home with family that trades privacy for peace of mind? In many cultures of the world, of course, aging parents live in the homes of their children in accordance with tradition; to live separately would be unthinkable. In more transient societies, where people move for jobs or related economic reasons, the aging of a parent or grandparent has a ripple effect on the rest of the family. Regardless of the specific physical circumstances, which have a big role in determining which features of life are possible and which are, or will become, intractable, the explicit or implicit debate among families confronting aging is about what can be done independently, by which many nondisabled people mean *by themselves*. Whether to stay in or leave their own homes, above all, but also smaller but weighty decisions, such as whether to travel alone or accompanied, whether or not to drive, in the daytime or evening—these choices carry big emotional weight, because acting "by ourselves" often stands for a set of assumptions about competence, functionality, and self-worth. It's the way most design for older adults—most people, period—gets framed: products and environments for independence as self-sufficiency.

But Kittay insists on an acknowledgment of basic dependency as a fact of human life that can be just as richly imagined as independence. There are closures and openings to this state of affairs as well:

> When we recognize that dependency is an aspect of what it is to be the sorts of beings we are, [then] we, as a society, can begin to confront our fear and loathing of dependency and with it, of disability. When we acknowledge how dependence on another saves us from isolation and provides the connections to another that [make] life worthwhile, we can start the process of embracing needed dependencies.

Kittay is a scholar, but she's not writing purely in the abstract; she is also the mother of an adult daughter, Sesha, who has complex cognitive and physical disabilities that will always require significant assistive care, so she has experienced these gifts firsthand:

I have received from my daughter Sesha a knowledge of, as [philosopher Alasdair] MacIntyre puts it, "the virtues of acknowledged dependency," and of the extraordinary possibilities inherent in relationships of care toward one who reciprocates, but not in the same coin; one who cannot be independent, but makes a gift of her joy and her love.

Redefining independence, in the legacy of the ILM and the technologically mediated life of a man like Steve, helps nondisabled people to reconsider their own tacitly held ideas about asking for help. But redefined *dependence*—as a plain fact of our lives that might also, in part, be salutary? This is its own profound lesson. Dependence, perhaps especially when it comes to cognitive disability like Sesha's, is its own kind of misfitting, because it always implicates far more than one person. Dependence creates relationships of necessary care—care that may be undertaken by individuals, families, local communities and municipal organizations, churches and mosques and temples, states or nations, or all of those in some mix. Kittay has understood from her own experience of acute dependency that her daughter's situation is distinctive, but it's also, in the perspective of any human life span, utterly common. "People do not spring up from the soil like mushrooms," Kittay writes. "People need to be cared for and nurtured throughout their lives by other people." In her relationship with Sesha, Kittay describes a mutuality more mysterious than transactional. Reciprocity that is "not in the same coin" forms an attachment that might look, from outside, asymmetrical: one party giving and the other receiving. But dwelling together, offering and accepting forms of help, is never a mechanical, zero-sum exchange.

Dependency and the care it requires may be the most distilled definition of disability and also the most universal. Some scholars claim that disability may well be "the fundamental aspect of human embodiment." *The* fundamental aspect? What a notion—that the universalizing experience of disability, states of dimensional dependence from our infancy through the end of life, might be the central fact of having a body, or rather *being* a body. It's an idea that could alter one's very sense of self, if we let it.

And yet, these scholars note, nondisabled people perpetually go to great intellectual and emotional lengths to distance their own bodies from the experience of disability in others: "The disabled body is imagined not as the universal consequence of living an embodied life, but rather as an alien condition." Steve's condition registers as deeply alien indeed to many people. There are philanthropic donors whose gifts, combined with Medicaid funding, make Saling House possible but who have never personally been on site to visit. They can't bear to see the people living there—to be in a room where bodily need is so vividly apparent.

Each year, Steve takes a trip with his two best friends from college—five days of debauchery, usually in Montreal. There was a third friend, whom he describes as "the alpha in the group who brought us all together"; he was the best man at Steve's wedding. But that friend has never made the trip with the group. It's too hard for him to accept his condition, Steve told me in an email. Steve has decided to be gracious about it; he says that prior to his own diagnosis, he didn't know any disabled people and might have acted the same way. "I had the same shallowness as this third friend, so I understand," he wrote to me. Still—is it so hard to connect Steve's condition to the fundamental human fact of disability? When Amanda came to my classroom, students could find their own bodies meeting the world, with assistance, on a continuum where she too lives. Steve, however—without any romanticism about it—is the truest cyborg, surely, that I know. He is integrated, body-to-machine and back: the body-*plus*, with its independence, its interdependence, and its dependence all made plainly, unavoidably visible. Two things can be true of his experience at once, and Steve taught me both: that he and everyone else would immediately, resoundingly rejoice in a medical cure for ALS, and that a life worth living, a life with both independence and dependence meted out in its material and immaterial goods, might also be built.

How much can be understood in rooms—rooms for voicing and signing our many languages with one another, for making household and bodily decisions, whether by ourselves or with our many modes of dependence on others? How can Gaston Bachelard's idea of *inhabiting* make our rooms into places where we might belong and build our lives, dwelling in places that embody "the essence of home"? The rooms in this chapter could never be described as "inert boxes," in Bachelard's term, because they have been so roundly *experienced*—by people in bodies that are out of sync with the majority culture, and yet experienced much more deeply than many people might imagine. But the rooms that make up the homes for Maya, or the generations of students after the Rolling Quads, or Steve and his housemates, are experienced not just in the intimate, domestic way we might automatically associate with residences. They're not just the inventions of a single person for their own enjoyment and belonging in a space; they are the products of collective imagination and labor. DeafSpace, and the Cowell Hospital dorm, and Saling House— they were each built from an urgent sense that being at home might be experienced in an expanded and altered form of body and well-being, and from gathering the will and assistance of others to see a new design realized. They ask us, too, to rethink all the deeply dependent features of our own homes, conjoined by walls and foundations or pipes and wires to the people and infrastructural systems—water, waste management, electrical power grids—that support our many bodies. Homes are never private palaces.

Architects and environmental psychologists sometimes use the term *action settings* to describe built spaces—in every way opposite to an inert box or a private palace. Action settings are not defined by an architectural style; they arise from a combination of our personal experience, our relationships to others who are present, the physical features of the envelope around us, and the patterns of activities that take place there. Action settings use all of these messages to help us make sense of how to behave in and use a space—a school or cathedral or public park—how to read all its cues and act accordingly with our bodies and voices. It's a more generative way to describe the world than assessing whether we "like" or "don't like" a building or a neighborhood. What is the setting for action that's available here? Which of those elements is cueing us in a satisfying way? Perhaps a DeafSpace dorm has little on the surface in common with the Cowell Hospital residence at Berkeley or with Saling House. But as action settings, these rooms immediately light up. With their inventive materials, shifted relationships, and adaptive structures, they embody ideas—and actions and interactions—of rooms as *home*, built around the facts of difference and disability. They ask each of us to reconsider our old ideas about independence and they ask us to seek with new eyes what it might mean to dwell.

STREET.

Geography and desire lines: Atypical minds and bodies navigate the landscape. Making space truly common.

The right to the city is like a cry and a demand.
—HENRI LEFEBVRE, *THE RIGHT TO THE CITY*

In collaboration with artist Wendy Jacob, Stephen spools out tape paths that break up the sensorily overwhelming cityscape of Boston with bright neon lines.

When I met Stephen, he was twenty years old and a longtime lover of maps, navigation, directions, and transport of all kinds. He had a job he loved at a popular tourist attraction in Boston—the "duck tours" that traverse both city streets and the Charles River on a combination truck-boat, an amphibious vehicle based on a World War II military model that crawls across the city throughout the warm months, lurching into the water at a climactic moment. Stephen is talkative, often gregarious. He's a talented piano player, speaks English and Italian, has an encyclopedic memory for dates and geography, and is also, as we say in the twenty-first century, on the autism spectrum.

I talked with Stephen in his family's living room about the hand-drawn map he'd sketched of the now-defunct A Line of the Boston metro. He had dozens of books on geography, the history of trains, and large-scale maps of the whole state of Massachusetts. His mother, an Italian-born legal scholar, told me that Stephen is the family's human GPS, reliably charting their daily paths through the city he's memorized. Stephen took on his mother's heavy accent when he spoke with her in English, as though switching into a personal dialect, with a precise ear for intonation and emphasis. Every time he addressed me, however, he returned seamlessly to the flat, bemused English that's typical of American adolescents.

I was visiting to talk to Stephen about the transformation that he'd been through over the previous decade. When he was ten, his interest in mapping and navigation was already deep and abiding. It was also profoundly hampered by fear. He'd been terrified of open spaces: plazas, beaches, any undifferentiated expanse too unwieldy for him to process into a coherent visual field. He craved sections and discrete division; he liked breaking the world into parts. For several years, he wore glasses with clear lenses in them, despite having perfect eyesight, because he preferred a compartmentalized visual frame. "I thought I needed a kind of focus then," he told me. He wanted a world bounded, contained. He found borders, lines, and edges reassuring: graphic lettering on signs, formal and informal paths. They made the world intelligible.

So when those boundaries and edges weren't available—as when a night's snowfall created a massive white field between the house and the car—Stephen's paralysis could be dramatic, even in the face of the need to get to school. His parents had been coached by teachers and therapists to break down challenging situations into small, legible parts, and they'd learned that linear guides were an effective tool for tackling open spaces. They would use sticks or other readily available materials to loosely configure lines in the field of white to mark a route, to forge a temporary way. Stephen could follow them as a guide and, with assistance, venture out—that is, when he absolutely had to. But he resisted most invitations to open expanses, landscape and cityscape alike. He and his mother sat indoors during more than one beach vacation, making geographical drawings instead of enjoying the sun and the length of shore. The lines and maps he loved weren't primarily a means to enable actual travel or even modest adventures in those years. They were the end itself: mental models of the world, organized and all-encompassing.

A shared interest in lines is what got Wendy Jacob, an artist and, at that time, a professor at MIT, talking with Stephen and his mother at a child's birthday party. MIT's faculty in the arts all have practices that broadly blend ideas in the sciences or mathematics into studio works—sculpture, performances, architecture, and more. Wendy was drawn to mathematical lines for their singular elegant dimension, clean and infinite, and she'd been staging performances with lines in a series of works called Between Spaces.

The work was realized first by suspending a steel cable—a giant, out-of-context tightrope—through an unusual location, and then having the line performed by its natural master: a tightrope walker. Wendy arranged one of those cables to cut through an ordinary family house, window to window. Another was hung at an odd juncture where MIT's library stacks adjoined an open, cavernous warehouse. Wendy worked with MIT engineers to calculate the loadbearing, and she engaged circus professionals to execute their precise skill in performing. Wendy thought of each of these performances as instantiations of a mathematical line: a uni-directional path, normally invisible, brought to life in these curious stages and performed without the expected drama of the circus. In mathematics, after all, lines are everywhere and never really end. "Theoretically," she told me, "they go on forever."

At the birthday party, Stephen had been stringing yarn between pieces of furniture, cutting through his own volumes of domestic space. Wendy, who had made artworks in collaboration with Temple Grandin, the autistic self-advocate, engineer, and writer, held an ongoing interest in autism spectrum conditions. Wendy wanted to know more about Stephen's linear modeling of his environment, and his parents were glad to find someone who shared his interests. An unusual friendship ensued.

Starting in the late summer of 2008, Stephen and Wendy met in her MIT studio, with open-ended plans to make something. "I would make things to explore in the building," he told me, "things like tunnels and maps." Soon their time was spent applying lines of blue painter's tape all over the white walls. They divided and subdivided the room, slicing its open space with adhesive lines. This practice continued until one day in the late fall when Stephen suddenly marked a solid shape on one of the walls, a large blue irregular mass. "It was like an old *Looney Tunes* cartoon," he said, remembering it, "where a character would paint a tunnel on the wall" and it would magically become an exit.

After that, Wendy suggested that they move the work outdoors, into and around the city, and the Explorers Club was officially born. It was a partnership that would take them all over the neighborhoods of Boston every Friday afternoon for two years, armed with pink or orange tape, the colors that mark off a construction site, spooling out lines and lines and lines through the city's pocket parks and public markets, through its bus stations and open lots. As a rule, they chose places where Stephen had never been: the far end of the blue subway line, all the way out to its last stop, Wonderland. The vast brick plaza of Boston's City Hall. Wendy and a research assistant would mark a path together, and then they'd invite Stephen to try, step by step, to walk out upon it. A small troupe of amateur geographers, armed with caution tape, they generously remapped the flat geometries that are everywhere in the built environment: the modest park lawns, the bricked commons. Slowly, over weeks and months, these lines became a shared mode of navigation for Stephen and the club of three—an unusual prosthesis that helped an autistic ten-year-old master public space. Aided by lines, Stephen learned to inch his way across swaths of cityscape.

Wendy has always said that the project wasn't a form of therapy or a teaching tool. It was a free affiliation around a shared enthusiasm with no measurements or outcomes attached, and no consideration of replicating the work elsewhere. But the Explorers Club, over two years, gave Stephen a critical tool: an unorthodox kind of compass, a way to make the ungainly world more manageable. Likely aided by some maturation in his sensory nervous system and the hours of practice he'd logged, he eventually took lines of tape and walked with confidence through other geographical unknowns, like a piazza in Bologna—the kind of vast space that had previously been impossible for him to comprehend and traverse.

In his living room a decade later, Stephen showed me a photograph of himself and his father floating in the Dead Sea, the body of water known for its saltiness that borders Jordan, Israel, and Palestine. Stephen is twelve. The salt gives the water a density that makes swimming an effortless act. "It was incredibly hot," he told me, both of us looking at the image of the two of them squinting in the sun. They are buoyant, weightless. There's a line of orange tape strung between them.

A utism lives in a brutal minefield of stereotypes: the savant, the obsessive, the socially distant autodidact. But in our time, the nature of autism has become a culturally contested matter. While many advocates seek funding for cures, genetic research, or behavioral therapies, others—vocal, activist, increasingly public others—roundly reject the very notion of autism as a disorder. As in nature, where biodiversity is a phenomenon of strength in multiplicity, the idea of neurodiversity, these advocates argue, invites us to see atypical minds and bodies as capable of making a powerful contribution to the world not in spite of but because of their ways of being.

The quest to know and understand autism has been an elusive business for a century. As journalist Steve Silberman shows in *Neurotribes: The Legacy of Autism and the Future of Neurodiversity*, the historical origins and myths of the autism diagnosis are only in recent decades coming into greater contextual clarity. Autism was first identified almost simultaneously by two different psychiatry researchers—Hans Asperger in Vienna and Leo Kanner in the United States—in the early twentieth century. In his research and writings, Asperger resisted a pathological approach to autism, convinced that the gifts and subtleties of this condition were worth attending to with respect and care. (Other historians have more recently shown evidence of his simultaneous complicity with the Nazi eugenics program that institutionalized many of these children, thereby guaranteeing their deaths by starvation or lethal injection while being officially reported as illness.) Asperger's 1944 research ultimately languished in obscurity after his clinic was destroyed. Meanwhile, Leo Kanner simultaneously identified autistic traits in a 1943 research paper, but he posited that autism was rare, and that its origins were likely caused by neglectful parents. The result was not only to pathologize autism as a disorder but to cast a set of medical diagnostics over it that were shameful and stigmatizing. It would take decades before the early disability rights movement, changes to the definition of autism in the *Diagnostic and Statistical Manual of Mental Disorders*, and legislation guaranteeing children more inclusive school environments helped autism become more widely understood as a developmental condition, and still longer before the notion of neurodiversity entered broader awareness.

Even with this new understanding, most research into the causes and nature of autism has framed it as either a medical matter to understand biologically, a pernicious threat to unsuspecting families via vaccinations or other environmental causes, or a political identity that should be respected and accepted without qualification. As Silberman shows, funding for autism research soared between 2000 and 2010, abetted by reports of an autism "epidemic" on the rise. In 2011, Autism Speaks, the largest, the most vocal, and arguably the most controversial advocacy organization for the condition in the United States, partnered with the Beijing Genomics Institute to fund an ambitious study that would sequence and analyze the complete genomes of ten thousand people who were part of families in which two or more members were on the spectrum. The study cost $50 million, and its promise was touted as "transformative." But after rigorous analysis of the data, the results were inconclusive at best. The genetic makeup of autistic people turns out to be all over the map—that is, just like the population in general. There were so few correlations that even the most commonly appearing genetic structure was present in only about one hundred of the ten thousand study subjects. The reality seems to be that autism is infinitely dynamic. After long, fierce debates, most researchers now understand that autism is a spectrum in the fullest sense. You might exhibit some traits but others not at all. You might occupy some trailing end, some feathery fringe of the spectrum that might be altogether undetectable by others. It's only now becoming clear just how common it is to occupy some location on that spectrum.

To be autistic is to be alive and alert to the world in ways that are both non-normative—when gauged by the standard operations of post-industrial cultures—and also pleasurable, challenging, adventurous, heartbreaking, and the whole gamut of qualities that attend any life. The disability of autism is an urgent political reality, to be sure, because most cultures aren't organized in a way that allows autistic people to easily flourish. But an increasing number of people on the spectrum would say it's not a disease they "suffer." There are lots of ways to be human. The question is whether the built world—its wearable gear and devices, its household products, its rooms and buildings and cityscapes, but also its social frameworks—can accommodate and mediate all the ways of human being, and whether the human-made environment can alter the weight of history by expanding and reshaping its inherited structures.

In Stephen's experience, a big open expanse wasn't just one of the features in his visual field; it was the overwhelming foreground. A blanket of snow or lengthy shoreline would take precedence over everything else, making the world noisy and confusing instead of a safe backdrop for an ordinary day. It's a form of what's been labeled Sensory Processing Disorder, a common challenge found in the cognitive profile of autism, and some of the prosthetics designed for it are things you've probably never heard of: body socks, pressure vests, rocking stools, chew toys. Objects like these are usually low-tech designs for "tuning" the sensory nervous system, what a physical therapist will sometimes call "getting organized."

With the assistance of prosthetic tools, *getting organized* is a process designed to help the body and mind attend to the "executive functioning" required for essential daily tasks. Sequential decision making, sustained focus, and linear problem solving are skills required to carry out even mundane operations like getting dressed and out the door in the morning. All can be hampered by dysregulation. Sensory processing anomalies run the gamut from hypersensitivity that can make a light caress feel like an assault, or the tag on a shirt a painful, irritating friction, to hyposensitivity, where the body underregisters cold, heat, or pain. For the hypersensitive, a stretchy Lycra vest may provide a calming deep pressure. For the hyposensitive, chewing a length of curly rubber—it looks like a loop of old-fashioned telephone cord, suspended from a lanyard around the neck—can provide the sensory input a body craves.

The engineering and design of these objects make up a growing sector of assistive and adaptive technologies. You can buy socks that lack the seam that usually runs along the toe line, irritating to some, or a length of stretchy elastic to suspend between the legs of a chair to quiet your bouncy, fidgeting feet. There are six-pound calming pillows for your lap, cocoonlike chairs for reading while nearly submerged in fabric, and all manner of squeeze devices for your hands, like a host of variations on rosary beads. These devices are for subtler uses, the less visible but nonetheless real misfits between our bodies and our constant negotiations as we move through private and public spaces, long waits and loud gatherings, large crowds and small groups.

Like the bright neon barricade tape that helped Stephen manage his sensory processing, the tools for tuning the sensory phenomena of the world come in all kinds of unexpected but also familiar forms. Even without a diagnosis of Sensory Processing Disorder, we navigate the built environment of industrialized life in hundreds of small ways, adjusting our bodies to the spaces we inhabit and the spaces, in turn, to our bodies. To quiet our nervous system, we wear earplugs, dig in the garden to relax, invite the sensory changes in our body that a run over pavement brings. Plenty of people who are not autistic would say they need certain physical conditions, and perhaps especially certain kinds of sensory input, to function well intellectually, to make creative choices, to feel calm and, yes, "organized." The state of the physical corpus is deeply tied up with the health of the brain; an overwhelming amount of research says so. We just have to find the right mix for our individual constitutions.

If a loud restaurant overwhelms you, the sound of pots and pans crashing sends you into a rage, or you find a roller coaster at breakneck speed to be oddly relaxing, then perhaps you recognize yourself on some spectrum or another. The sensory world is a weird mix for human bodies, each of us moving around in a singular flesh envelope. We traverse the rooms and streets we like and we hate, turning the little knobs of the world's sensations by our choices of what to wear, how to walk, what to include and what to shut out. Augmented by headphones or a hat pulled low, decked in seamless socks or technical Lycra or fuzzy sheepskin, wrapped tightly or loosely, rocking or fidgeting or chewing our nails—each body makes a stream of conscious and unconscious choices, knitting together a habitable personal universe minute by minute by minute. When I see Stephen mapping the world in lines, breaking its space to tame it, all I can think is how perceptively he's externalized an invisible but fundamentally human need: to build bridges that temporarily edit the shapes, or sounds, or sights of the world. And I wonder: Who else is looking for lines, a little lost in space? Who else is seeking a way?

In any sizable park or green space, you'll likely find two kinds of paths: the formal kind, paved with brick or concrete, and the informal kind, the paths made by people walking over and over a stretch of grass, wearing away the green and carving a scruffy emergent line in its place. These are paths made by sheer repetitive use; they're not anyone's executive decision but arise one choice at a time, collected in aggregate. Most of us know them as friendly disobedience: they're shortcuts, maybe, or just the most commonsense pathway from one frequented site to another. Urban planners call these paths "desire lines," or sometimes "cow paths," "pirate paths," or the slightly stuffier "counter-grid trajectories." They indicate yearning, some planners say—either to have formal paved lines where there are none or to actively carve out a different path where one had been prescribed.

Desire lines can provide low-tech crowdsourcing for urban planners and architects, letting the habits of the walkers dictate to them how a space is best traversed, rather than trying to decide it up front. Some large campuses (in the United States, examples include Michigan State and the National Institutes of Health) have postponed the paving of pathways until desire lines have first been created. One celebrated case was in the remodeling of the Illinois Institute of Technology, a project taken up by Dutch architect Rem Koolhaas and his firm, the Office of Metropolitan Architecture (OMA) in the late 1990s. That campus presented a particular conundrum for achieving social cohesion. It had doubled the footprint of the original institution but had only half the enrollment. What kind of building would energize and unite the school? The OMA group studied desire lines and used them to plan the campus center, unified by a long single roof. The building wasn't so much a new creation as an observation of extant use: it effectively enclosed the pathways and connections between activities on campus that were already established. The single-plane building is like an archive, capturing activities in motion. It took its form from travel behaviors made newly visible, not from a series of architectural types pre-identified for recreation, shopping, and the like.

This kind of practice is a human-centered design approach to landscape, paying close attention to the details of movement and patiently observing an area over time. Urban planners can imagine the lines of use among people by looking at the evidence of where they go but also where they stop, looking for accumulations of litter or cigarette butts to see the natural pauses in walking routes. These traces of human use are handy for tackling traffic congestion, for example, or addressing pedestrian safety.

There are other kinds of desire lines. Bicycle advocates in many U.S. cities, unhappy with existing provisions for their safety, have found ways to mark off bike lanes guerrilla-style, with tape or paint laid down in the wee hours, and sometimes with (of all things) toilet plungers, which are cheap, easy to cover in reflective tape, durable, unexpected, and impossible to miss, even when you're flying by in a car at forty miles an hour. Bikers place a line of plungers to highlight a bike lane that's getting ignored or one they want to see formalized. In Wichita, Kansas, a set of plungers along a bike lane got wide press coverage, and the city reinforced the lines with official infrastructure "flex hit" posts that perform in a similar fashion two weeks later. The plunger lines are an example of what's often called *tactical urbanism*—citizen work to create more desirable cities with small experiments that have an air of performance about them. They're public actions undertaken grassroots style, intended to be photographed and distributed widely via social media. In this case, they create "what if" possibilities by marking new desire lines for safer streets.

But desire lines may also be evidence of something more than pure practicality. The casual disobedience of a desire path as an alternative to the formally prescribed walkway is remarkable simply as a human choice, willfully out of step with the way things are. Cities and towns are often planned, well, by *planners*, by people tasked with creating systems that make mathematical sense for groups at the scale of hundreds or thousands. They roll out pathways conceived around the efficiencies of use and cost-benefit, shunting people up and down stairs or elevators, nudging them between turnstiles and onto trains for the fastest transport. And these efficiencies are often to the good. But the emergent, informal, human-made lines are organized not only by efficiency but by *desire*. The human individual is also making a path through life, through interiors and exteriors, a life that cannot be measured in abstract bureaucratic terms.

At twenty, Stephen no longer uses fake glasses to frame the world, and he no longer uses tape to get around. The photographs from the Explorers Club's adventures, the tape zigging and zagging in front and behind, are all that's left of the time they spent together, three huddled geographers and the occasional guest explorer. They are the relic images of Stephen's own paths, generated from both discomfort and daring.

What archive could ever fully house the evidence of *desire* in the millions of walkers who daily traverse a cityscape, all the wishing and wanting that drives each path, with all their untold forms of assistance, all the getting "organized" that got them out the door and into the street? "Walking is a mode of making the world as well as being in it," writes Rebecca Solnit in *Wanderlust*. The book was published in 2001, before most of us could imagine carrying around tiny GPS tracers in our pockets and handbags, the little machines mapping our steps made of obligation and steps made of desire. But even those lines are ghosts and not really our possessions, are they? The Explorers Club has its lasting evidence.

The social life of city sidewalks is precisely that they are public," wrote journalist and activist Jane Jacobs. In her landmark book, *The Death and Life of Great American Cities*, Jacobs was defending the importance of busy street life: what happens in dense, informally grown neighborhoods where people live and work and shop, all in the same area. When all of life happens in proximity, streets are both heterogeneously social and safer, wrote Jacobs, creating a naturally occurring form of support for the strangers who share it. "Lowly, unpurposeful, and random as they appear," she wrote, "sidewalk contacts are the small change from which a city's wealth of public life may grow." But you can only see and be seen, only get into and out of the shared public life of the world, if you can get down the sidewalk in the first place.

The need for accessible streets and sidewalks has brought about visible changes to the contemporary cityscape, and the most profound change is also the most modest: the *curb cuts* that you'll find now at many street corners in cities all over the world. They're nothing but edited concrete shapes, an inclined plane newly cut and laid where there had once been a step down from the curb into the street and back up again on its opposite corner. Instead of this up-down motion of the stair-step transition, you now travel in a glissando of diagonal drift from the sidewalk to the street and its inverse on the other side. The angle of the "cut" is calibrated for ease of use by people in wheelchairs. It needs to be shallow enough for safe passage when someone is traveling alone, whether using their hands to push their wheels or a motorized chair commanded by button or switch.

If you've ever dragged your wheeled luggage over a dozen blocks of busy sidewalk, walked your bike through a crowded area, or pushed a sleeping baby in a stroller in a gingerly fashion, hoping not to wake her with sudden bumps and jolts, then you stand in this inheritance, too: a commonplace rite (and right) of passage that was rolled out at improbable scale, taking old cities and chopping their corners, one by one, bending their angles for the work of wheels.

Curb cuts are retrofitted alterations to the textures of the built environment, and they make plain a global history of city design that has largely been planned around a normative "user" of its streets. Design makes possible or impossible the means of practical use, but it is also a pointed commentary on the meaning of bodies that move through spaces. A city with only hard-angle ups-and-downs, curbs, and steps to all its doors and entrances is a city that assumes a strong ambulatory body, unencumbered by injury of any kind, unaccompanied by an older adult companion or a young child in arms or holding a hand while learning to walk. Streets, that is, have long been designed for working men with physically strong bodies and no meaningful caretaking responsibilities—no obligations to parents or ailing siblings or offspring—not the kinds of caretaking that create a clunky, inefficient, assistance-borne passage through city streets to get to the places a body needs to go. The revolution in street corners seems like an obvious civic good now, a common-sense softening at the edges of the built environment, a simple solution to buffer the concrete shape of a world built with homogenous users in mind. But it would not have happened without disability activists' long, hard fight.

Long before the Americans with Disabilities Act of 1990 mandated curb cuts at all street corners, disabled people had pointed to the design of the street as a key locus of their political rights—the sidewalk that stands for being in public space, and therefore in the public sphere. Disabled people had called for curb cuts since the 1940s as a form of rehabilitation for veterans, requests that went largely unheeded except for rare exploratory programs. Throughout the 1950s and 1960s, a sustained movement of disability activists at the University of Illinois at Urbana-Champaign worked to create accessible housing and transportation for students with physical disabilities there, including placing planks of wood over stairs to loudly call attention to inaccessible buildings, lobbying for curb cuts in the local campus area, and more. In the late 1960s, disability activists in Berkeley, California, including Center for Independent Living leader Ed Roberts and his circle, mixed and poured concrete to smooth the passage from sidewalk to street in a DIY "editing" of the city. It was a hacker-style approach to creating the conditions of the future. Others took a more forceful tactic, smashing the concrete curb away to create the cut required in guerrilla-style street action, even if the resulting plane would be rough-edged. You can see the remnants of this form of design protest in the Smithsonian today—a preserved bit of sidewalk taken out in 1978 by disability activists in Denver who were smashing away curbs as an unignorable public gesture, one that was both pragmatic and pointedly expressive. It's a hunk of concrete that stands for so much: the insistence that access would not just await new architecture but must be created by unmaking and remaking the inherited street itself.

Disability scholar Aimi Hamraie recounts this history as an example of disabled people redesigning their own worlds, efforts they took on themselves as a call to action for their own civil rights, without the paternalism of rehabilitation. In the present day, curb cuts are so common, both ordinary and even mundane, that most people know nothing of this history. But the resistance to their widespread implementation was protracted and fierce. Outside a few small communities like Berkeley, where vocal activists won some local implementation, there was little understanding of the chicken-or-egg problem of accessible design. "When we first talked to legislators about the issue, they told us: 'Curb cuts, why do you need curb cuts? We never see people with disabilities out on the street. Who is going to use them?'" recalled Roberts. "They didn't understand that their reasoning was circular."

The fight culminated in a dramatic day of protest in March 1990. At the "Capitol Crawl," people using wheelchairs, leg braces, and canes made their way to the hundred steps in front of the U.S. Capitol building in Washington, D.C. Then they began to climb those stairs, leaving behind whatever gear couldn't come with them, using their arms or whatever body parts they had available for mobility. Children as young as ten participated in what became a very public, strategic spectacle. That protest is considered by historians to have been the tipping point; the Americans with Disabilities Act was passed in 1990, guaranteeing curb cut changes at every city sidewalk corner and ramped entrances at all newly constructed buildings, among other new provisions.

When sidewalks and streets are built for some bodies and not for others at the scale of infrastructure, they create what political scientists Clarissa Rile Hayward and Todd Swanstrom call "thick injustice"— inequities within urban structures that are "deep and densely concentrated, as well as opaque and relatively intractable." These segregated structures of the hardscape appear to be permanent, anonymous, and inevitable; it's hard to imagine their conditions otherwise. When urban planning sends investment toward rich neighborhoods for redevelopment amenities like public parks and advantageous zoning laws, inequities result. Urban planning decisions like these have long compounded racial inequities in many American cities, for example. These injustices are meted out over long periods of time, with public and private sector decisions mixed together and therefore hard to disentangle. Hayward and Swanstrom argue that when injustice is tied up with the physical spaces of cities and the policies that create them, it becomes "difficult to assign responsibility for it—and hence difficult to change." That's why the curb cut's history—a successful "editing" of the built environment that arrived via mandate of federal law—is so impressive and historically profound. The careful reshaping of the sidewalk allowed more kinds of bodies on the street but also into public life. "When disabled people enact politics," writes Hamraie, "they also design and build new worlds."

Curb cuts aren't the only ways city systems have been edited to make way for all kinds of bodies. On a "kneeling" bus, a mechanism lowers the vehicle's front right corner by letting air out of its suspension system, taking the entrance platform from 13.5 to 9 inches off the ground and making the buses easier to board for older adults and people using wheelchairs and other adaptive gear. Buses were retrofitted with this kneeling equipment in the United States starting in the mid-1970s, also as an initiative of disability activism. Moreover, many global cities have stickers on bus and subway windows reminding riders to offer the most accessible seats to fellow travelers with disabilities or others who might need them. On a bus in the Basque country of Spain, I once spied a large sticker on the window that depicted eight or nine human icons in all kinds of situations that would warrant seat preference: an older adult with a cane, a pregnant woman, a mother with a young child, a wheelchair user, an older adult with a young child, an entire range of scenarios. I couldn't read the text, but it didn't matter, because the images said it all: *Use your head, and take note: those around you may need that spot to rest more than you do.*

The very solidity of the built world gives it stubborn staying power. The guarantee of infrastructure—its reassuring layout as a networked system—can turn into thick injustice when planned for abstractly determined human populations, without consideration for a variety of actual bodies. Rectifying forms of thick injustice doesn't come easily; assigning accountability and making legislative change is the work of decades. When I'm up and down on the curbs in my own city now, tackling the stairs or taking the elevator, observing the kneeling bus as it lowers for easier entrance and watching for who sits and who stands on public transportation, I think of those protesters on the steps of the Capitol building, insisting on a wholesale reshaping of cities and demanding a legally binding promise to make it so, one edit at a time.

Meanwhile, desire lines of all kinds proliferate: the unofficial paths carved out by hundreds of feet and wheels; the proto-lanes for bikes that arrive unofficially overnight; the guerrilla-style smoothing of curbs with concrete, cutting along the diagonal for freshly minted informal ramps; and a ten-year-old's lines of tape. Big changes take small experiments, public spectacle, pilot programs, testing, and patience. Who makes the city? Its systems are not easily mutable, and yet: people do make and re-make its passages and textures. The public street stands for the public sphere. It's the ordinary hardscape built for use, a means for getting from here to there, *and* it acts as a symbolic platform, a way to arrive in public, present and accounted for.

O n a blustery day in November, I got off the train in a Dutch town called Weesp and made my way through a series of residential neighborhoods to the entrance of De Hogeweyk, a planned village. After checking in at a welcome desk, I was let through a set of doors that opened onto the village's main plaza, where streets, dwellings, and businesses appeared to converge. The weather was gray but not frigid, and people were out and about. Trees and potted plants surrounded sets of tables and chairs on the expansive pavement, and from the main entrance, two streets took my eye in different directions, both lined with small storefronts flanked by bright signage and built of handsome brick. At the intersection were street signs that read "Boulevard" on one corner and "Theaterplein" on the other.

The signs—like the stores, the streets, and the plaza itself—were both real and not. That is, they marked geographic places in De Hogeweyk, and the businesses were, in a sense, businesses. But unlike an ordinary streetscape, this one was disconnected from the larger city outside its border: streets, plaza, and the rest of it lay inside this locked facility, which is home to approximately 150 residents who had come to this makeshift town to live out the last years of their lives. Each has dementia advanced enough to warrant full-time care. And De Hogeweyk, with its gym and its hair salon and its restaurant, is the opposite of a nursing home. Residents are free to wander there, and so was I. Once I'd left the doors behind, I quickly forgot the heavily fortified security enclosing its perimeter.

Some of the people who passed me looked physically fit and elegantly dressed. By their initial behavior and appearance, you wouldn't have known they were patients here. Some walked; others used wheelchairs.

Some were unaccompanied, and others were engaged in conversation. A woman I took to be a visitor had a pair of dogs with her and accompanied a resident in a wheelchair who seemed to be silently enjoying the wind on her face. Others, more advanced in their conditions, lay nearly motionless as they were pushed, covered in blankets to stay warm.

I had an appointment with a woman named Iris, who works full time as a communications liaison at De Hogeweyk, hosting visitors like me nearly every day of the year. I had wanted to get a feel for the place first, so I had made a reservation for lunch at the restaurant and arrived early. The restaurant turned out to represent much of what makes De Hogeweyk distinctive. It was open both to residents and to the general public. Staff and customers seemed to share an unstated agreement to make polite and sensitive room for residents who entered. Some were there with family members, but there was also a cluster of colleagues having a celebratory business lunch. I sat near a young mother with a baby napping in a stroller and a toddler she kept having to chase. My Dutch is decent, a holdover from a long-ago graduate school interest in Dutch history, and I could follow the conversations around me fairly well. The mood was relaxed, like at any quiet business-casual restaurant: cloth napkins, jazz playing in the background. A floor-to-ceiling bar at one end, lit from behind with multicolored lights, showed off the bottles of alcohol available to residents and guests alike.

At one point a man, mild-mannered and hunched, gray-haired and using a wheelchair, entered the restaurant at a slow pace. He had a small portable radio on his lap. It was clear that he was not looking to sit down and eat; he had simply wandered in. At De Hogeweyk, wandering is part of life. He made his way among the tables, looking to quietly engage the patrons. The bigger table of celebrants politely ignored him, but the young mother spoke to him as though his presence were expected and welcome, modeling for her child a mode of interaction that was both warm and respectful. The waitress greeted him like an old friend.

Iris joined me at the end of my meal and told me a bit about her own and De Hogeweyk's history. Now in her thirties, she had been trained and had worked in the hospitality field. But she'd grown tired of the slick commercialism of the hotel business and sought out an experience where she would make relationships with older adults. Her new job is "different, but not so different" from hospitality work, she said, and over the course of the afternoon, I started to understand what she meant. As with most young Dutch people, her English was precise and enunciated; she spoke in a rapid-fire, energetic lilt. "People come from all over the world," she told me—"from students, to governments, to organizations"—to witness a new idea in memory care that's utterly unlike the typical hospital ward or nursing home. "But the design is not leading," she said, when I expressed my interest in the architecture and streets. The campus plan reflected and reinforced a series of values, she told me: "The vision came first."

De Hogeweyk had been a standard nursing home on this same site in the 1980s, built with the expected long hallways of rooms, with centralized nursing stations and the general feel of a hospital. But in the 1990s, De Hogeweyk's board members, several of whom have backgrounds in care work, started seeing their own family members aging and in need of assistance, and they had an honest reckoning with the status quo of their own caregiving for residents. Would they want to place their own parents here? Or see themselves as residents, at some point in the future? The answer was no, and this realization set them to reinventing memory care, to make it much more like life in general. They implemented shared cooking tasks and community activities; they planned enriched and active programming for residents, to simulate the continuities of pre–nursing home life as closely as possible. They eventually articulated six pillars that outlined their values, including the obvious category of "Health" but also "Life's Pleasures," a commitment to making available fresh air, church attendance if desired, concerts, a pub, and the like.

"Favorable Surroundings" was foundational to the organization even before the opportunity arose to work with architects on a brand-new site, paid for mostly by the state. In 2009, the De Hogeweyk board commissioned Molenaar&Bol&vanDillen architects to build a new environment that would fully realize this core value. The details were crucial: in addition to the simulated village, they wanted private bedrooms in dwellings that would house small groups of residents, no more than six or seven to a dwelling, decor available in a series of styles to be familiar to different residents, kitchens and laundry facilities in each home. The way the architecture of the place expresses the six pillars is akin to the way Deaf-Space evolved—from a set of precepts that became physical.

All the De Hogeweyk buildings are constructed at human scale, just one or two stories high. When you walk down the boulevard, there's a fitness center with all the requisite equipment in its sunny interior; a sandwich board out front advertises upcoming classes. The music center's walls are covered in images of instruments, and vintage horns are laid like sculpture on the piano against a wall. The theater hosts musical performances and movies for residents, and it also rents out the space for events that aren't intended for the community, as a supplemental stream of income. And this is the Netherlands, so there are bikes: ingenious tandem styles, side-by-side instead of front-and-back, making a wide sturdy base for easy balance. The shops and stores operate as distinct activity centers for the residents, the kind they'd otherwise experience in one big multipurpose room at an ordinary nursing home. Each of the distinct village elements strongly broadcasts its function by the material choices that go into its structure. All the visual elements—colors and shapes and signage—are selected to minimize confusion as the residents find their way around town, to say vividly and reassuringly, *You are here.*

The interiors of the living quarters are decorated in a range of styles, determined by an opinion poll that encompassed a wide range of national representation, and are matched broadly to residents' backgrounds. They range from relative formality, with fixtures like chandeliers and ornate furniture, to much more simple arrangements of wooden chests of drawers and cushy recliners. Meals are planned, shopped for, and cooked in the fully operative kitchens by the residents, according to their abilities and interests, with the help of a care worker. There are knives and other sharp implements in the drawers. The small potential risks in safety are considered worth the gains in autonomy and agency, Iris told me, adding that visitors like me are often astonished at this setup. "The Americans are always amazed," she said to me, pointedly.

Some 47 million people worldwide live with dementia, and that number is growing in places where populations are disproportionately aging overall; projected numbers by 2050 are around 132 million. When people think of dementia they think of memory loss, especially short-term memory. But the confusion about navigation that attends this condition is as important as the loss of recall. Getting around, and having the seams of the world more or less patched together in a coherent whole, becomes challenging when the plaque and tangles that attack brain cells do their degenerative damage. Getting oriented and organized for a walk becomes difficult, and the confusion produces anxiety, and the anxiety creates agitation, and agitation, if it's bad enough, often gets treated with pharmacology. But psychotropic drugs like Haldol, while they may quell agitation, also often render patients passive.

At De Hogeweyk, the use of these drugs has dropped dramatically in the twenty years since its pillars reshaped its community life. Rather than combat agitation, the environment is structured to maintain orientation and calm in the first place, to ward off that cascade of fear. There's a trust in design at work here, in the structure of the sensory environment, arranged just so, to do at least some of the work of treatment. The surroundings have been constructed to flex to the needs of the body, instead of the other way around. And while the commitment to "favorable sur-roundings" is far more vibrant there than anywhere else I've ever seen, many nursing homes have been influenced by the Green House Project, an effort to deinstitutionalize elder care in Western cultures, including making more homelike designed environments for residents with or with-out memory loss. And where degenerative brain disease is the leading diagnosis, design is increasingly being recognized as a key component of treatment.

Lost in Space: Architecture for Dementia is a smart and elegant coffee-table-style book about the burgeoning field of design for the conditions and symptoms that come with dementia. It reflects ideas that clinicians and caregivers alike have generated for creating surroundings to support both navigation and memory. Some of these choices are relatively elemen-tary and are increasingly common in assisted living facilities and nursing homes. The use of warm, organic paint colors like greens and blues for interiors reassures patients with homelike visual signals. Memory boxes can be hung like artwork in patients' rooms, little time capsules that carry familiar objects and materials from a patient's past.

Some solutions are more specific to dementia and are increasingly be-ing implemented in memory care centers. Elaborately painted murals over exit doors help counter patients' attempts to escape. High-contrast paint colors help them discern entrances and doors or to distinguish be-tween dishes and placemats at a table. De Hogeweyk takes it a step fur-ther, creating by a careful design an entire community that looks and operates like a well-signposted version of the real world but is in part a simulacrum. The grocery store is stocked with real food arranged in aisles, appearing in every way like the ordinary street version. The same goes for the pub and restaurant and hair salon. Residents experience them as authentic. All the apparatus of the nursing home—the medical files and monitoring of treatment—is invisible; it's "backstage," as Iris said, invoking a theatrical metaphor that's commonly used by staff.

Iris and I were taking a walk around the campus when we were stopped by a woman who seemed to be on her own brisk jaunt, unattended. She wore a quilted pink jacket against the wind, her pale cheeks reddened by the outing and short silver hair moving in the wind, her greeting hearty and convivial about the winter weather. She'd paused to ask us directions to the town of Hilversum, clearly sure it was just around the corner. Iris invited her to walk with us instead, but after some moments deliberating, the woman opted to continue by herself, leaving us behind with a purposeful stride.

I pressed Iris on her easy manner and assured responses in the face of the woman's evident confusion: How do you decide whether to play along with someone's fiction or to gently convey that the truth is not what they think? I wondered how and why she would ignore a fabricated story that was taking a resident away from the here and now rather than insist that she return to the real. For some dementia patients, surely, the transience of a sense of logic and time must be the most discomfiting feature of their lives. Why not guide with a more deliberate hand? Iris said this was an art form she had had to learn. "I tend to believe it's better to tell the truth," she said. "But every person needs a different approach." She thought it the responsibility of the staff, including people like herself, without direct training in care, to make a careful judgment call in each encounter. If it seemed harsh to push a reminder of reality on a resident in a given situation, it was wiser to refrain. "You have to investigate a bit what you say and don't say. It's different for everyone, and then in two months it might be different again."

Farther on our walk, we caught up to the woman. In the meantime, another woman had joined us. This second woman, too, would have passed my notice as a patient; she too wore street clothes and was walking with an apparent sense of purpose that passed for ordinary. But as we walked on with the woman in the pink jacket, this newcomer became visibly agitated, for reasons that were unclear to me—although I could well imagine how irritating it would be to have a slippery grasp on the real, and to be caught in a series of conversations and encounters that seem to change on a dime. At the moment, her ire seemed to be directed at the woman in the pink jacket. Iris handled the situation masterfully. She gently herded the four of us into the restaurant where I'd eaten lunch, calming the frustrated woman with repeated gentle requests to join us for a cup of tea. At first the woman insisted on standing instead of sitting; she argued briefly with the other woman on a matter I couldn't make out. Finally she relented and sat; the waitress brought the tea. Her face softened, her shoulders relaxed, and she appeared, for the moment, to have conceded to the present. In real time, Iris had used the physical space, the passage of time, and the surroundings themselves to de-escalate a moment of fear-borne aggression.

We four settled into ordinary talk. Passing the milk and sugar and reaching far into the dustbin of my memory for Dutch vocabulary, I turned to the woman in the pink jacket, the traveler to Hilversum, and asked her how long she'd been at De Hogeweyk. "Oh, I don't live here," she assured me. "I'm just renting a room."

The mixed approach to reality and fiction that De Hogeweyk embodies is a big matter of debate in the field. Simulated surroundings do more than just offer navigation help; they also build a remembered or known world that doesn't really exist. The falsehood presents a conundrum to family members and professional caregivers alike. Researchers and clinicians all over the world employ approaches that take a widely varying stance toward truth and reassuring falsehoods. Some believe that affirming patients' narratives more effectively encourages their happiness. Others maintain that telling the truth, even when it's difficult to face—reminding a patient of the long-ago death of a spouse, for example—is the only ethical choice when you're building long-term trust. These are serious questions, and disagreement in the field continues. But for the staff I met at De Hogeweyk, the "performance" is rationalized as a preservation of continuity, a connection to ordinary life in the face of change. It's architectural treatment, you might say, intended to mitigate the coldness of clinical logic that marks institutional care.

After I parted from Iris, I sat in the courtyard for a little while to consider all I'd seen and heard. Life in the street—as a means of access or a means of desire, as a way to knit together the past with the slippery present—requires its structure to be both practical for our many needs, visible and invisible, and sometimes a kind of stage. I thought about how the street at De Hogeweyk is ingenious most of all for being *semi*-public space. Its structures are necessarily segregated but not sealed off from view. Its model village combines simulacra and real working parts, like the restaurant that faces two ways, built for residents receiving memory care and for residents of the town nearby. What might it mean to design streets or buildings or neighborhoods where more of the commons could be shared—where life in public, physically and therefore socially, could open and close in this porous way?

It's easy to think in terms of opposing goods, either/or choices for our lives when misfit conditions arrive: in the case of dementia care, either a segregated facility or "aging in place" at home. But De Hogeweyk's mixed model for shared life is a way of thinking beyond either/or. Design includes unmaking and remaking with a combination of small gestures and big ideas, as each of the stories here shows. Stephen's temporary prosthesis made of tape lasted as long as it was needed. Curb cuts have lightly edited every sidewalk corner rather than replace the structures altogether—a subtle change that changes everything. And a village-as-medical-facility houses a restaurant that's partially open, partially closed. De Hogeweyk hosts visitors from around the world every day who marvel at this simple and powerful idea: that care might be creatively designed. All of these choices make the street more flexibly open, more of the time—by thinking through each of its features with deep imagination and pragmatic collective will.

The winter evening was falling precipitously fast. As I gathered my things, I noticed a woman nearby, pleading at the doors—the doors to the "real" world—to get out. With my attention elsewhere for the day, they had faded from my vision, but suddenly here was this reminder of De Hogeweyk as a nursing home. The scene wasn't cause for alarm; I gathered it was commonplace. Staff members were doing their best work to redirect her. But her distress and confusion were evident. I found myself thinking again about *action settings*, like those that arise in the buildings at Gallaudet and the Saling House residence. This time, the action setting came alive not just for the indoor space but for the hardscape, in the makeshift roads of the village. I hoped this woman would turn and locate herself again under the trees in the plaza, that she'd spot the warmth of window lights from the dwellings above coming on, see the restaurant gearing up for evening business, take in the smells of dinner. I hoped the street-stage might help set up her actions, send her its cues in a reassuring way. But it felt rude to watch at length, so I went through the doors and out into the night.

CLOCK.

Life on crip time. When the clock is the keeper of our days, what pace of life is fast enough?

A Singaporean pedestrian places an ID card enabled with "Green Man +" features over an electronic reader, temporarily extending the time the crosswalk signal will allow him to traverse the street.

f you're patient enough not to jaywalk, you know that at busy street intersections in many cities of the world, there are buttons to press that trigger signals on a screen in a box, usually mounted on a pole on the opposite corner. When the stoplight turns red, halting traffic, the screen displays the icon of a walking human, often in green LED lighting; below it, a set of numbers, usually bright red, counts down. The number usually starts at a set clock, indicating how many seconds pedestrians have to cross to the other side of the street in relative safety, before the impatient waiting drivers are free to step on the gas again. If the numbers are already getting down to the single digits and the street is wide, you might decide to wait if you've just reached the intersection, or to hurry up if you're halfway across. The allotted seconds are chosen by urban planners, and the number is based on some mathematical calculations: How busy is the intersection? And how wide is the given street? These two factors, plus one more, determine a crossing time, and that last factor is this: What's the pace of a normative human walking gait? How long does it take the average human to get from one side to the other?

The average is the standard, but a body is not an average, and in Singapore City, that nation's capital, an overall aging population has made the misfit condition of too-short crosswalk times more dramatically visible in recent decades. Singapore isn't alone in the new demographic reality of aging: most nations around the world report an expected growth in the percentage of adults over age sixty in their populations by 2050.

These changes play out in all kinds of material ways, but one of the subtler ones is the need for extra time. There are groups of people who just need a few more seconds—older adults, whose walking has slowed with age, but there have always been pedestrians with disabilities, perhaps using a wheelchair or some other assistive gear, or people living with chronic pain that slows one's pace. How do urban planners design for those extra seconds? How do old cities flex their structures for new conditions, especially when the conditions are the intangibles like time?

There are ways to do this relatively simply, of course—by unilaterally extending the crossing times at all signals, a kind of "universal" design measure that would slow the pace of every intersection. That kind of change would permanently foreground the experience of human walking, or wheeling, while also making it easier to travel in the company of small children. But it would likely slow traffic times all over town. It's also possible to extend the times in some crosswalks and not others. In New York City, crosswalk times at some intersections have been permanently lengthened in the past decade. These changes were demographic decisions, localized and limited in implementation—an acknowledgment of aging pedestrians in a given area and a planning decision to meet those citizens' needs.

What planners in Singapore City decided on was a nimble intervention in crosswalk infrastructure. On hundreds of the intersection boxes where normally you would push a button to call for the pedestrian signal for your passage, there's now an electronic sensor that will read a discounted transit card that's issued to all senior citizens in Singapore. That card is now additionally outfitted with a technology that is called the Green Man + (Plus) program, a reference to the "green man" who appears midstride on the countdown screen. When it hovers over the sensor, the card activates an extension to the intersection's crossing time—that's the "plus" part: three to twelve more seconds, depending on the width of the street. After the card-holding user has crossed with the extra time, the signal reverts to its ordinary allotment, making passage both accessible and efficient. Pedestrians with disabilities who have not yet reached senior citizen standing can also apply for a Green Man Plus card, gaining time for wheelchair use or some other qualifying condition. As of 2014, after half a dozen years of pilot testing, there were five hundred intersections with this technology in Singapore. It's a loosening of the city structure's very fabric, and not in the typical manner of adding benches to bus stops, or directing traffic with rotary structures and better signage. Those are urban designs for moving optimally through features of physical space. The Green Man Plus card is urban design for something that's equal parts elusive and fundamental—a way to make the structures of time expand and contract, on command.

T he medical field has a long tradition of describing disability in reference to time," writes disability scholar Alison Kafer. "'Chronic' fatigue, 'intermittent' symptoms, and 'constant' pain are each ways . . . of describ[ing] disability in terms of duration. Time is everywhere in disability language, writes Kafer: "'Frequency,' 'incidence,' 'occurrence,' 'relapse,' 'remission': these too are the time frames of symptoms, illness, and disease.'" How long, at what speed or slowness, how often, when it started, what the future prognosis will look like—these are the temporal words doctors possess and use for talking with patients about what's happening in their bodies, whether describing the slow creep of change that happens with age or puzzling out a new and mysterious set of symptoms.

The words *acquired*, *congenital*, and *developmental* start to get more abstract, with *developmental* doing even more complex work, Kafer observes. A condition may be developmental in that it extends over the life span, but the term may also be used to describe another kind of friction with respect to time: a "developmental delay" to describe a child or adolescent always has a standardized and normal idea of individual progress as its background. How long should it take for a child to learn to crawl, or begin to speak, or read independently? To occupy the tail ends of the bell curve is either to be precocious—that is, early and therefore advanced—or to be delayed and therefore behind. Kafer notes the temporal speech that arises in even our most casual questions to one another about our bodies: Were you born this way? How soon will you recover? The differences and deviations in disability, short-term and long-term, physical or cognitive, are unquestioned references to the apparent fixity of time. Kafer notes that the universal nature of disability itself is premised on the clock: "Disability studies' well-rehearsed mantra—whether by illness, age, or accident, all of us will live with disability at some point in our lives—encapsulates this notion, suggesting that becoming disabled is 'only a matter of time.'"

Among disabled people there's a bigger catch-all term, a slang for this particular mismatch: it's called life on "crip time." *Crip* is short for *cripple*, a name that disabled people have repurposed in an act of political reincarnation, dropping the degradation attached to a word that was used to describe them in the past, *cripple*, and investing it with in-group pride. "Crip time" is flexible shorthand in disability culture, used to indicate a range of uneasy relationships to the pace of contemporary industrialized life, with its relentless and clock-driven organization of hours and days. As a disabled person, to say you're "on crip time" on a Tuesday might signal the extra time that it takes you to get to the train platform or in and out of a public bathroom. It can also stand for bigger systemic fits and starts—the spiky, unpredictable time it might take a person to proceed through a fairly rigid K–12 education that's built on all kinds of normative chronologies.

Crip time can also be "time travel," writes disability scholar Ellen Samuels—forward and backward motion through ages and stages brought on by misfit conditions. Samuels was in her early twenties when her body changed swiftly and dramatically with disabling illness, and her doctors had no real answers. The dissonance between her age and her bodily state caused her clinicians to speak of her life as unnaturally skipping in time: "You've lost so many things already in your life: your parents, your health, your independence. You have a level of loss we would usually expect to see in someone in their seventies."

Samuels writes that she went from someone "with health problems" to "*being* a problem, apparently insolvable." To be so young and so beset by physical illness ejected her from any ordinary timeline, and she knows she's not alone in this state:

> Disability and illness have the power to extract us from linear, progressive time with its normative life stages and cast us into a wormhole of backward and forward acceleration, jerky stops and starts, tedious intervals and abrupt endings. Some of us contend with the impairments of old age while still young; some of us are treated like children no matter how old we get. The medical language of illness tries to reimpose the linear, speaking in terms of the chronic, the progressive, and the terminal, of relapses and stages. But we who occupy the bodies of crip time know that we are never linear, and we rage silently—or not so silently—at the calm straightforwardness of those who live in the sheltered space of normative time.

Crip time, then, can describe an out-of-sync plotting of life in its mismatched state. But the term can wield a pointed critical bite, too. How long does it take, or *should* it take, for a body to move through the world, the forty-plus-hour workweek, the demands of caregiving for children or ailing parents, the daily commute of the body with its changing needs over the life span—a pregnant body, an aging one, a body in recovery after a bad injury? And then there are situations of crisis, conditions that alter both our bodies and the work we do as a measure of our worth. Is the clock of industrial time built for bodies at all?

The pace set in contemporary schools and workplaces presumes a form of able-bodied productivity, an ideal of speed and efficiency. That's not a literal clock but the one in the background, invisible yet always ticking. It's a mismatch for all kinds of people, because it's not simply a timepiece. It's an economic instrument. The body and mind are a combined organism with infinite complexity, but the measures at hand are those of the clock: whether and how the social-cognitive mind and body of a person is checking off the milestones for a given set of skills in a timely way. It makes us obedient to the demands of competition—for getting through reading levels and standardized tests and job interviews by "getting ahead," by getting through faster than others, becoming compliant with the unquestioned ideals of productive *worker* time—worker *time* established for an encompassing identity of worker *beings*. Many of us tend to accept the expectations of the industrious clock, with its minutes ticking away and announced to us everywhere, as fixed and permanent.

But crip time announces the possibility of another way of thinking, outside economic time. Crip time is not just a call for the values of family-friendly schedules or work-life balance, important as those may be. "Crip time is flex time not just expanded but exploded," writes Kafer—exploded by the raw disruption of disability, the atypical body or mind that refuses to be the insolvable problem, and indicating instead a larger social and political mismatch at hand. The extended crosswalks in Singapore are the faintest signals of a bigger design confrontation with time, a misfit scenario that may be the most diffuse and intractable of all. A disabled person, caught up at odds with the mechanized and economic tempo of life, isn't merely apologizing for lateness. That person is diagnosing not the slowness of a body but the unyielding, ossified containers for quick and efficient *productivity*—the expected time that life will take, life for everyone.

It's the crip time of *delay* that organized my family's life when our son, Graham, arrived with a diagnosis of Down syndrome in 2006. We were immediately the recipients of state-sponsored "early intervention" in his infancy—professional therapy sessions to stimulate his gross-motor and fine-motor development, his speech and sensory intake, some of it carried out in our home and some in clinic settings. These sessions were the supplemental kinds of playtime meant to compensate for his predicted lateness in all things babyhood: rolling over and sitting up, making sense of speech and communication, and more. Early intervention is a recent design in the United States; since 1986, it has been offered by most states, a set of services designed to maximize the potential of infants and toddlers before they enter the school system. In the main, they're a wonder of support and advice descending on young families in a vulnerable stage.

Our experience in those early months included lots of ordinary first-baby life, of course. My husband, Brian, and I, animated by a swoony, terrifying, animal love for a baby we'd just met, took in the hard-wired perfume of his newborn smell, misplaced his dozens of miniature socks, compared stroller models. We had to figure out how to help him sleep and change his diapers with the steep learning curve of newness that always comes with the first go-round, all attended by worry and obsession over getting it right, just like most parents everywhere. We walked every day in the canyon parks of Los Angeles, this baby strapped to one of our chests or backs. We were mostly buoyant, happy, scolding our errant dogs and enjoying the trees and making plans with our friends. Much of it was ordinary, at least from the middle distance.

But some of it was decidedly unordinary, too. We had a surprising postnatal diagnosis; the pregnancy had been totally normal. We got a false negative on a genetics blood test, and I had routine ultrasounds and all the standard check-ins without concern. We fully understood that only amniocentesis was a sure bet for certain diagnostics, but we opted out of that procedure. We weren't comfortable with the invasive risks it imposed, knowing we wouldn't terminate a pregnancy over Down syndrome or similar conditions. Graham had a beautiful birth at forty weeks, a late-morning arrival on a seventy-degree January day in Los Angeles. A couple of days later our midwives suggested, with so much kindness and sensitivity, that we get him tested, and our pediatrician delivered the news two weeks later. The news punctured the swell of our insides, permanently altering the happy discovery of his early newborn days. There's no other way to put it: on a Tuesday he was a singular human, the likes of which the planet had never seen before, and on Wednesday, somehow, he transformed in the eyes of others into a type, a *kind* of person now defined by statistical risks and telltale physical features.

Was it harder or easier this way? To hold your child in your arms, as smitten and sleep-deprived as you were the day before, while learning something quite new about him? Or to plan ahead for him in the abstract? We only know our own way, of course. It's only now that I can see the dissonance that was the most painful: to Brian and me and our close circle, he was still the much-wanted individual person he'd been since arrival. But to nearly everyone else, he became the diagnosis—forever described and understood and interpreted primarily by genetic status.

Down syndrome in its overwhelmingly common form is the presence of an extra chromosome that has no etiology of disease. It comes with risk factors for physical complications, some of them serious, but its phenotypic expression in physical traits is wildly varied—wildly varied, that is, much like every human life in its complicated sack of flesh. But significant delays in many forms are nonetheless guaranteed. In those early days, we were encouraged to interpret his tracking along milestones in two parallel sets of statistics—the arcs of time for average development in neurotypical children, and the alternate times for children with Down syndrome, also averaged into their own charts. That was the invisible, persistent tick-tock of passing time, with its markers of weeks and months acting as evidence: how he was doing ("for a kid with Down syndrome"), how we were doing at our job in caring for him (advocating enough?

flaming swords or bleeding hearts?), all made palpable by charts and measurements in a temporal frame.

Perhaps because we'd never parented before, we studied these charts religiously, and our families and friends also asked us well-meaning, pointed questions about Graham's skills and traits. They wanted to know—and we did too, hard as it was to admit—whether you could tell if he was more "high functioning" or "low functioning," in the common-place language of our day.

High and *low* are spatial words. They're a way to rank humans, you could say, in assessments that arrive relatively up or down, intended eu-phemistically to sum up a report about performance that is actually about time. How quickly was Graham doing infantlike things, and how nor-mally fast was he getting them done? The logic is pernicious, sublimated, and sweeping in its interpretation: If he walked on the "early" side, the thinking went, did that mean he would be "high functioning," and there-fore he would also understand, say, addition and subtraction on the early side? And did that also mean he might understand *all* abstract concepts in schooling in a timely fashion, his functioning attended by only modest delays? Would all of *that* therefore mean he'd have an ordinary and dig-nified job in adulthood, be less dependent on other people because he would be generally, overall, somehow, *quicker*? Meaning—quicker and therefore more normal, more worthy of respect? The language of high or low is an unavoidably global and totalizing assessment of achievement, and therefore, implicitly, a judgment. How "high functioning" one may be is a commentary on worth—how high a person's value will turn out to be in the eyes of the wider culture.

Graham sat up at nine months but didn't crawl at all. He didn't walk until three and a half years old, and didn't meaningfully run until he was eight. He spoke words at twelve months but not complex sentences until third grade. He'd memorized the alphabet at two—strong visual memory is often a strength in Down syndrome—but his handwritten letters, at thirteen, look like those of a very young child. The neatness of the bell curves for milestones among typically developing children, how quick or how slow, even with some fuzziness around the edges—these were never generalizable or predictive for him, not ever. Most timelines just didn't apply. And more important, his relative quickness or slowness compared to his peers and two younger siblings has thus far been of little interest to him as a source of self-worth. His association with schooling and extra-curricular activities like dance or sports has been primarily one of curios-ity and friendship—not *simple*, mind you, just overall *joyous*, uncolored by comparative percentages and reading levels. These days he wants to show us the slide presentations he's putting together for seventh-grade science class—he takes pride in it, and he also knows his neurotypical classmates are doing other kinds of assignments at their desks nearby. (We're all aware—Graham himself, parents, and teaching team a dozen strong—that the gaps between those assignments are growing wider.) Graham still looks forward to school and associates it with success. He also employs the occasional shrewd tactic to procrastinate or gently derail a teacher when he wants to, asking for a lot of breaks when the subject is especially chal-lenging, like math. In short: he's a recognizable teenager, doing lots of teenager things, but he's not playing the comparison game. His enjoyment in school is enhanced, not threatened, by the faster-paced skills of others.

This basic equilibrium in his sense of self may well change in adult-hood; we've watched our friends and acquaintances with Down syndrome struggle through their adolescent roles and relationships in mercurial pat-terns that are something like those of their nondisabled peers. But the in-sistent, clock-driven measuring of his childhood comes from others, not from him. Those assessments arrive by others *about* him, the terms and gradations uttered in the theater of the doctor's office and the classroom teacher conference.

Still—we didn't have the solace of hindsight when he was so young. We have lost countless, possibly needless hours of worry asking *when*.

And at every point in his life, people have asked us, or sometimes announced *to* us, where Graham lives in this conspicuous and maybe totally arbitrary standard of timing that results in a label of "high" or "low." The reasons and metrics vary, depending on context, but the spatial measurement of high or low isn't really about heights or depths. It's always about time. What people want to know, when they ask about "functioning" in an unconsciously roundabout way and covered in so much friendly goodwill, is whether and how much he is *retarded*. Most people don't dare use that language anymore, because it's considered rude and out of date. But high or low is actually about fast and slow. *Retarded* is ordinarily a neutral and descriptive term about slowness in time. But it says so much—not about Graham but about the bald truth of contemporary life! How strange and yet how telling: slowness can so easily exit the realm of pure description and transform into its popular use as an insult, a clock word weaponized to diminish a human being.

Living with questions about time—about Graham and the diagnostics of his delays, but more urgently, the unknowns about his future— was the loneliest feature of our early months. Was it just families like ours who were facing down the clock, relentlessly (and resentfully) allowing it to measure our lives? It took me a while to see that our tiny family unit was connected to ideas far bigger than ourselves. It was more readily apparent for us upon Graham's arrival, but the clock measures productivity for everyone now, in a manner so internalized it feels all but natural to our bodies.

History shows that the early uses of clocks arise from both familiar and unexpected sources. Mechanical clocks, before setting up the rigidity of industrial time, made an early appearance in the monasteries of Europe. They were employed by men seeking the order of their daily "offices"—not the physical workplace offices of our own day, but

WHAT CAN A BODY DO?

the liturgical offices, the rituals of prayer designed for marking the early-morning, noontime, evening, and sleeping hours. For some forty thousand Benedictine monks in the thirteenth and fourteenth centuries, obedience to time was liberation by practice and rule, marked by clocks before they were commonplace, before there were clock towers, for instance, marking official city times. Lewis Mumford, in his classic history of technology, *Technics and Civilization*, reminds us of the social and emotional work of this early clock, keeping a desirable life set apart by spiritual habits that were measured by marking the hours. "Within the walls of the monastery was sanctuary," he writes. "Under the rule of order, surprise and doubt and caprice and irregularity were put at bay." Clocks marked the rhythms of sacred time for men who deliberately chose life outside the mainstream, and, in so doing, paradoxically helped create the normalcy of timekeeping that would overtake the entire industrializing world. Mumford includes this history to remind his original readers in 1934 of a truth that needs retelling in our own time: that tools and technologies both follow, and lead, and follow again from the ideals that cultures value. Monasteries, full of men whose life's work would never be economically efficient, "helped to give human enterprise the regular collective beat and rhythm of the machine; for the clock is not merely a means of keeping track of the hours, but of the synchronizing of the actions of men."

Synchronizing, of course, took on a life of its own in the centuries that followed. Clock towers, and eventually home clocks and watches, made it possible to carry this instrument everywhere and to measure the hours of everyone by its instructions. The clock made it possible not just to mark the passing of minutes but to imagine time as preceding other metrics for life. "Abstract time became the new medium of existence," writes Mumford: "one ate, not upon feeling hungry, but when prompted by the clock; one slept, not when one was tired, but when the clock sanctioned it." For centralized industry to work—organizing factories and farms, of course, but then dramatically, suddenly, with the vast expansion of the railroads—time had to exist outside locality. Standardized time, as agreed upon in the United States and Canada and rolled out on November 18, 1883, established a national agreement about time that echoed trends elsewhere in the world: the industry clock was king. The *Indianapolis Sentinel* reported the day's implications with a grin you can hear, but the prophecy also makes its way through:

> Railroad time is to be the time of the future. The Sun is no longer to boss the job. People—55,000,000 of them—must eat, sleep, and work as well as travel by railroad time. It is a revolt, a rebellion. The sun will be requested to rise and set by railroad time. The planets must, in the future, make their circuits by such timetables as railroad magnates arrange.

Industrial time is useful! We can hardly imagine life without it. But outsourcing authority to the mechanical clock has severed time from all connection to the body or the season by breaking life into a series of discrete units—the effect of which, writes Mumford, "helped [to] *create the belief* in an independent world of mathematically measurable sequences." Mathematics, that is, soon wed to the directives of industry.

It's this matter of belief that is so consequential, because even the twenty-four-hour day, examined empirically, is an average at best. There's nothing about the rotation of the Earth, the organization of the monthly calendar, or the real expanse of longer and shorter seasonal days that's remotely as precise as the mechanical clock. Its units are mathematical, but its origins are cultural. And the clock will certainly never represent the strange experience of time in our *lives*—the ways that it is fundamentally interactive and ephemeral, the time travel of memory and expectation. Physicist Carlo Rovelli writes that the nature of time is so hard to characterize that it's more accurate to call it an event rather than a structure, "more like a kiss than a stone." All physicists will tell you the same: that time is a slippery thing appearing every day in the guise of the exacting clock, unable to yield its measures for us, "waiting for no one" and all the rest. Each of us has to decide, insofar as we can, whether to let the economic clock dictate our minutes and hours, and how we'll know if we measure up.

But where does that leave, in Alison Kafer's words, a future for crips? Looking back can help us look forward.

D esign for intellectual disability in my own country, the United States, and much of Europe has been powerfully shaped by a significant architectural structure: the institution, or asylum, which arose as a novel method for centrally housing and managing people with cognitive disabilities starting in the first half of the nineteenth century. The designed segregation of the asylum for intellectually disabled people (along with the asylum for people with mental illness, in parallel) was largely a response to industrialization. With families more transient, a new service economy emerging, and a more distinct separation of home and workplace, the care of family members whose capacities hindered the efficiency and centrality of the nuclear family became untenable for some, necessitating formal state intervention for care. For what we now call intellectual disability and was then called "idiocy," there were both social and economic reasons for segregation.

Prior to the nineteenth century, intellectually disabled people would have either remained within family structures or been under the care of almshouses, a community resource that brought a mix of both vulnerability and protection. But once census data of the mid-nineteenth century showed the prevalence of intellectually disabled people in almshouses or prisons in aggregate, suddenly those numbers appeared threatening as a social problem. Fears about intellectual disability coincided, too, with general fears about cultural "degeneration" that plagued the Victorian era: the squalor and poverty in newly crowded cities and the applied understanding of Darwin's natural selection in a social and political frame, with social "fitness" newly articulated as an achievement. A new association of "idiocy" with pathology necessitated and raised the stakes for more centralized care. "Idiocy" was recast from its status as a relatively benign pity to a more serious social threat, linking intellectual deficiencies to both physical and/or moral deficiencies, especially in poorer families. Historian James Trent, in *Inventing the Feeble Mind: A History of Intellectual Disability in the United States*, writes that this association between disability and poverty was key. It was the *economic* vulnerability of these children and adults, a vulnerability that extended to their families, that made them "feeble," indistinguishable from their status as *unproductive*. As a group, then, "idiots" were increasingly managed by custodial supervision in a designed campus for care, outside the locality of families and communities. And *management* is indeed the key term here. Trent notes that the rehabilitative educational mission that many institutions claimed did not always result in returning "feeble-minded" children to their communities. Instead, they created extended "custodial" working environments for adults *within* institutions that handily perpetuated their own institutional necessity. Institutional care, which between roughly 1850 and 1950 was sometimes benevolently rationalized as management, sometimes born of fear, became in every case bound up with and even indistinguishable from control.

From there, things got worse before they got better. The advent of eugenics and its policy ideals for shaping healthy nations spelled out horrors for cognitively disabled people, among many others, in the United States and especially under Nazi rule in Europe. It was a nationalist-spirited campaign to modify the future of the country by discouraging reproduction among "unfit" citizens in the hopes of populating the world with those deemed fit: racially or ethnically pure, intellectually and physically healthy, and sexually normative. For people with intellectual disabilities and psychiatric conditions, the eugenics era included brutal abuses of power: mass sterilization, medical experimentation, literature for parents that encouraged euthanizing disabled infants, lethal injections in institutional settings, all in the name of safeguarding and promoting a healthy citizenry. In parallel with the "oralism" push in deaf education in the same period and the "ugly laws" prohibiting "unsightly (disabled) beggars" on the public streets of cities, the eugenics movement at the turn of the century warned about all manner of "defective" and "degenerate" bodies. Above all, the organizing principle of eugenics had its roots in disability not exclusively as an identity but as an encompassing idea: that "unfitness" of many kinds could be measured, known, and eliminated, and that the mismatch between bodies and the world could threaten so-called human progress dramatically enough to warrant eradication.

Trent recounts the rise and fall of the institution, including its lowest point in the eugenic era, and reports the loose and overlapping sweep of terms for intellectual disability used in the last two centuries. The words demonstrate a change in meaning over time: "a disorder of the senses, a moral flaw, a medical disease, a mental deficiency, a menace to the social fabric, mental retardation, and finally a disability." The terms bear out the inevitable entanglement between a history of medical diagnostics and the social and political ideas that shaped those diagnostics. But there's been "so much change, and so little progress," writes Trent:

> At various points in their history, these nouns began to be qualified: *defectives* became *mental defectives*, *imbeciles* became *high-grade* and *low-grade imbeciles*, *morons* became the *higher-functioning mentally retarded*. Later, the *mentally retarded* have become *mentally retarded persons* and *persons with mental retardation* and, in some circles, *persons with developmental disabilities*, and *persons specially challenged*. In 2005, the American Association on

Intellectual and Developmental Disabilities agreed to change the label to *intellectual disability*.

Lots of parents like me have an instinct to celebrate this apparently happy evolution in language, but Trent knows better: in this process of linguistic change, he writes, "*essence* has apparently been liberated from *existence*, [and] *being* from descriptions of it. Behind these awkward new phrases, however, the gaze we turn on those we label intellectually disabled continues to be informed by the long history of condescension, suspicion, exclusion, and pity." And the language, he adds, may well be a convenient cover: "While our contemporary phrases appear more benign, too often we use them to hide from the offense in ways that the old terms did not permit."

The segregated institution is no longer the standard of care for people with intellectual disabilities. As with the de-institutionalization movement for people with mental illness, the 1960s and 1970s brought about the reversal of institutional logic, slowly shutting down most of these warehouse-style residences all over the world. Instead of living life sequestered and nominally safe, the thinking went, people with developmental and cognitive disabilities should be included in schools and communities—in mainstream classrooms and in "group home" environments for sharing life skills, with needs for the attendant-designed supports that these moves imply. But for all the change in terminology, all the relocation of bodies to spaces outside the institution, all the well-meaning legislation in the world, the statistics are still hard to take: the employment prospects for disabled people are grim, and people like Graham are at much higher risk for abuse and neglect. "In a society that defines and confines all meaning and worth in terms of production, profit, and pervasive greed, intellectually disabled people will likely be exploited," Trent writes. The settings and locations for inclusion, he shows, have only so much influence, while the powerful critique offered by crip time is ignored.

What is a happy life? What would flourishing look like, for people like Graham living on the most palpably constrained forms of crip time, and for any of us living in the shadow of the industrial clock? For people with intellectual disabilities, Trent writes, the *means* of care—segregated or inclusive classrooms, institutions or group home residences—has so often been collapsed to become the unquestioned end in itself, because the larger assumption remains unexamined. Economic productivity—a life performed in normative, regulated time—is still the unquestioned and overwhelmingly dominant metric for human worth. This means-and-ends matter is a central question for all of us: What future goods should we hope for, for ourselves and the children in our lives, and what are the mechanisms that will foster those goods? At the dinner table with my own family, in talking with the parents of my local community, and as a professor with my high-achieving students, I'm constantly thinking about the stories people tell me about their ideas of the good life, the presumed ends our efforts are intended to foster. How much do we organize around our academic or professional training, versus our social and behavioral growth and our wider connections to our communities? What components make for a "good school" for our children? Test scores? The accelerated-in-time and therefore "advanced" classes? The economic tempo of the clock shapes our every conversation.

It has been Graham's singular creativity, his adaptive workarounds, his enigmatic learning curves that have wrested my husband and me from the grip of rigid time. It has been our relationships, with him and near him, that have invited us to reconsider the general organization of parenting around milestones and percentages and test scores for all three of our dimensional, changing children, and instead to keep returning to our children themselves, their bodies meeting their respective worlds. Graham helped break us of any instincts to make him into a scripted success-story project—an all-too-common and benign-seeming determination to beneficently "prove" his worth by hoping that he might be exemplary among children with Down syndrome, a certain kind of achiever that other people would understand. This is a negotiation every family must work out with clear-eyed assessment: Who is it the child wants to be? And where might the advocacy efforts of parents proceed, wittingly or unwittingly, from what are actually *parental* needs—perhaps prescribed ideas about familial success and happiness—rather than from radical acceptance and support?

Instead, with time, we have to come to consider independence, and dependence, and interdependence with much more robust imagination. Whether Graham will live on his own with regular scaffolded supports, or with a group of friends, or with us for the long term, we have learned to recognize and to counter our culture's repulsion about the plain facts of dependence and assistance. My husband and I have learned to acknowledge our own "derivative dependence" that comes with being any parent at all—our needs for childcare, schooling, parks and infrastructure that exceed our capacities—and to acknowledge Graham's economic and existential vulnerability that also shapes *our* lives for the long term. But that recognition doesn't leave us isolated. On the contrary—we now understand that we are part of a big story about the unfinished work of the built world. We have deepened our connections, extending our political concerns beyond our nuclear family, and we have tried to help organize more desirable futures, ones where disability might be publicly visible. Graham's invitation to life on crip time, with the clocks of our lives expanded and exploded, has beckoned us to a kind of freedom. It has never been and never will be simple. But it has offered us a deep repository of learning, a lens for thinking through all of our memberships in the human family.

The discovery of crip time, as a term and a generative idea, gave me the gift that theoretical language can bestow when it steers clear of pure jargon: it handed me a description of life that I intuited as a parent, but for which I hadn't found formal expression until I'd gone deep into the disability scholarship that had been lively and productive long before Graham's birth, ideas articulated by disabled people for decades. Crip time was one of many ideas in disability that, once I'd witnessed and lived them in our interdependent domestic life and then sought out their history among disability's many thinkers, landed my son and our entire family in a restless venture that far exceeded the bounds of our home life. Never mind the noise of popular culture, with its stories of personal "victory" over disability, all the spotlight on individuals who bravely go on "despite" their body's shortcomings. Crip time is something else entirely. It suggests that the clock may be every bit as much the culprit in the mismatch between a life like Graham's and the world, not simply his genetic status. I'm no longer lonely in the face of economic time, because I see it clearly now, exactly as it is, a monolith that encompasses all of us: a mechanized, utilitarian, ultimately impoverished scale for human value. My son doesn't need a gentle and pacifying form of "inclusion." Inclusion is necessary, but it will never be sufficient. He needs a world with a robust countervailing understanding of personhood and contribution and community in it, human values that are alive and operational outside the logic of the market and its insistent clock. He needs it, and so do the rest of us.

That's the series of design clues I'm looking for now. Will it be in products, digital or material, or furniture, or rooms to help build a future for developmental disability? Will it be in material goods at all? When I think about Graham, I consider all the possible design interventions that might bridge the misfit between his body and the built world. I think about Cindy's archive of tools, sitting at the ready in her handbag, the OXO peeler in my kitchen drawer and the closed captioning on my television, and the hundreds of cardboard chairs lining the Adaptive Design workshop. I think about the Signing Starbucks, about the elegance of its simple moves to rearrange the architecture and technology and thereby change the relationships that unfold there. But how do you design for the conditions of crip time? Some challenges defy prosthetics. There are lots of good ideas out there, but it will take more than extra crosswalk seconds to design a future with slowness in it.

I've seen the glimmers of good ideas in lots of places—designed services that arise in situations where furniture or apps would never suffice. Creatively supported employment in the post-institutional age for people like Graham is in short supply, but there are points of light. In the last couple of years alone, I've seen a bakery in Scotland, a tea shop in the Berkshires, a studio for handmade crafts in Ahmedabad, and a Texas ice cream parlor all run with and by people who have intellectual disabilities. I've met cognitively disabled people working as ticket-takers in museums and as assistants in early childhood classrooms. I've seen self-defense classes designed and taught to young disabled adults on the cusp of high school graduation. They learn verbal and physical skills for staying safe on the bus or subway, making their entrance to the workplace more secure. I've witnessed the growth of hybrid college programs for young adults with cognitive disabilities in the last two decades. These are ingenious curriculum designs, offering a combination of coursework and life skills practice, with training in budget management and cooking and more. The creativity is endless and necessary. I have eagerly sought out these possibilities, as a parent and a researcher, always looking for evidence of others seeking to imagine adulthood in a way I can see for Graham. But the best-designed service I've seen was a situation where people with cognitive disabilities weren't solely on the receiving end of the work of others. These disabled young adults were on the giving end instead, doing the building and rebuilding themselves.

A t a park in Dorchester, a neighborhood in Boston, a group of nearly thirty young adults met on a Saturday morning for a day of park cleanup—raking and picking up trash, neatening up the shrubs and grasses that border the ball fields and playgrounds. It was a colder morning than expected in early May, blustery and bright near the waterfront, and the group assembled near a picnic table in twos and threes. Some of us were wishing for an extra coat, having been overoptimistic about springtime in New England, and others clutched paper coffee cups. The city provided trash bags and tools to the group. What these young people offer is their community service, a form of civic volunteerism. It was a social and convivial atmosphere for many hands making light work, and everyone had come equipped with the clothing and gear to make it possible: sweatshirts and jeans, sturdy sneakers or hiking boots.

These young people commuted in from all the various parts of the city of Boston, by bus and by train, some using walkers or wheelchairs, others practicing their skills with memory, the better to navigate on public transportation. We rocked on our heels and pulled our shirt cuffs down over our hands, fidgeting to get warmer while we talked. The group included Josiah, confident and outspoken, leaning on a walker and returning for his fourth season; Travis, who was just starting out as a new volunteer; Marisa, a wheelchair user and a co-architect of the program since her own participation as a student in the very first season, who was finishing her undergraduate degree at the University of Massachusetts. There was Rose, tall and with a musically high voice, who narrated immediately the subway route she'd taken, noting how she nearly got on a train heading in the wrong direction. Sheelah, another first-timer, reminded her that the security officials in red shirts working on the platforms of the T—the Boston subway—would help her if she got lost. Some lengthy discussion ensued about the puzzling nature of the visual arrows in the maze of underground transportation: sometimes arrows on signage pointing up mean "straight ahead," and sometimes they mean "slightly to the right," and still other times, they mean "go up" the stairs or elevator. It's confusing, said Rose; she wanted to talk about the details, tell the story again, figure it out in the company of others. This is all part of the process of this volunteer opportunity for these young people—the ones who have been the recipients of so much assistance for so much of their lives, the ones who have been so little trusted with real responsibility. On this day, the roles would be reversed: they would do manual labor for several hours, adapted to their capacities. They'd rake or gather leaves, pick up trash or hold the bags for others. They were caring for common infrastructure, and they were doing it all in public.

I was with the sixth cohort of young people to spend a year as a "service warrior," or community volunteer, as part of EPIC: Empowering People for Inclusive Communities. These volunteers spend two Saturday mornings each month giving their time in one of two ways: making physical improvements to the city of greater Boston, whether it's in a park cleanup like the one I took part in, or prepping meals for a soup kitchen, or working in a community garden. On other Saturdays, they'd meet and talk about their futures—what self-advocacy looks like, basic disability rights and history, and rehearsals for prospective job interviews. They'd complete one hundred hours of combined time like this in a year.

Once most service warriors had arrived in Dorchester, everyone gathered in small groups for their first meeting of the day. Luke, a recent high school graduate from the suburb of Malden who had a job in afterschool childcare, is a graduate of EPIC and had taken on a paid staff role in this cycle as a team leader. Luke had a notebook with clearly written instructions and steps for the day in big type on each page. He held out the notebook and led the group through a series of call-and-response affirmations that the group always recites before the day gets moving. Some exchanges were about expectations—"Phones off / We on!"—and some were exhortations to claim control over their ways of being in public: "Show up / Show out!" And most telling of all was a recitation that reminded everyone to resist being on the passive end of what others decide. These words weren't just rehearsals for the day; they were rehearsals for life. EPIC wants to create a habit of taking charge for young people who've too often been conditioned to accept the agenda that others are setting—the care that turns into control. Each Saturday that they're out in public, they say these words too: "My needs / I express." That is to say: *My needs are mine, and I can't wait around for other people to notice what I need. It's my responsibility to speak; I can't let others speak for me.* And after the huddle, they set out to clean the park.

I n most contemporary schools and recreational activities, students with physical disabilities would not be automatically grouped together with students who have intellectual disabilities, and sometimes that's for good reason: their needs and wishes can be quite distinct. Some of these young people had accessed special education services for help with standard school curricula, and some had used accommodations for physical access only. But the program was set up to address a larger barrier that all of these young people faced, regardless of how they had done academically. They shared the task of countering the assumptions of so many people in their daily lives—that they were somehow in need of extraordinary accommodations just to participate in the structures of adulthood. EPIC is also mixed for a very strategic purpose: these young people need to see each other not as the *same* but as a cohort that is politically *connected*, a microcosm within the larger community of disability rights whose coalition has formed to bring disability—all kinds of disability—to the foreground. Who is the world built for, and also: Who does the building work itself?

EPIC's design—building capacity with service work, boosting the résumés of these young people with legitimate work hours, jobs training, and self-awareness—it all comes right at the moment that parents like me lie awake and worry about: young adulthood, when so much funding and support for people with disabilities stops. When Graham grows up, he will also need a world designed with some porousness and flexibility around its edges, and he won't be the only one. When people with disabilities are winsome children who are all possibility, under construction, with their stories on the far horizon, the structures of special education and adaptive sports and summer camps all rush in to make room. But what happens when that all winds down?

For service warriors who have physical disabilities, EPIC is a volunteer opportunity that will enhance their job prospects, as it would for any young person. But for the warriors with intellectual disabilities, the significance is amplified. "Special education" is commonly framed and (mis) understood as a one-way street: misfit students exclusively on the receiving end of assistance. In childhood, with the promise of their whole lives ahead of them, they are the subjects of academic tracking in schools. But in adolescence, their progress and rights to services start to converge with an eye toward an abstract notion of "job readiness," and so little imagination has been brought to the possibilities for these students' futures. They can be taught any combination of academic content and "life skills" in a classroom setting, but when the official stage of transition hits, at age seventeen, the questions get harder. What kinds of jobs will even be made available to them? Will they be safe? How will they ask for help they might need and still demonstrate their value as employees?

This group—young people like Josiah and Travis and Rose—needs an arrangement of services, curricula, relationships, and other intangibles that create pathways to enter not just public buildings or public streets, but the public sphere itself—life in the mainstream, for people who've been taught that their differences are the most remarkable thing about them. As their historical counterparts who cut curbs, and turned hospitals into dorms, and crawled on the Capitol steps, these young people are building new worlds, too—the ones they want to be part of. On Saturdays with EPIC, a life-altering role reversal arrives for the people least likely to have ever been called up for leadership.

Jeff Lafata-Hernandez is EPIC's creator and leader. Working for City Year in Boston, he realized that the powerful service experience it offered—a prestigious line on a résumé, a close-up look at civic life and public infrastructure—wasn't available to young adults with disabilities. Tall and reedy thin, Jeff leads with a soft-spoken demeanor that can mask the radical nature of the work he's doing—his insistence on respect and dignity in all interactions with the young people he trains. Those priorities are the engine of his organization, and it shows up on service days. "Sometimes a passerby will approach us and talk to one of the service warriors in that telltale baby voice—you know, patronizing with cute talk?" Jeff told me. "*How-are-youuuu-today?* That kind of thing. And we'll talk about it afterwards." He notices this kind of behavior immediately, but with a service warrior, he'll dispassionately observe and gently repeat what just happened. He'll ask: "Is that okay with you?" For these young people, too often the tiniest successes are celebrated; real challenges are rarely offered, and potential is rarely tapped. Jeff has been working so long with this population that he knows two things: that this kind of sentimentality is rampant in the exchanges these young people have in public, and that they have also gotten used to it.

On a weekday in between service weekends, I met the group when they convened at a suburban office park, where a health insurance company had agreed to partner with EPIC to run mock job interviews. Staff at the company volunteered their time at lunch to sit with service warriors and practice for the job interviews that would soon be part of their lives. We made our way past the lobby and through corridors with the expected generic beige and brown walls, furniture, and chairs, finding our way to a basement event room in the process of setup—tables arranged with chairs opposite one another for the practice-run interviews, while others against a wall were lined with boxed sandwiches and bags of chips.

Before the interviews began, the EPIC group gathered to sit and talk among themselves about what kinds of questions are appropriate in an interview situation and what kinds of questions they might be asked about their own experiences. Prompted by Jeff and other EPIC staff, the group considered what counts as job "experience" and soon realized that their work with EPIC itself would be a strong line on a résumé. They talked about when and whether it was legal to ask about accommodations for their disabilities in the workplace, and about how to conduct themselves in conversation. Soon the men and women from the corporation, the role-playing "employers," started to trickle in for lunch and took their seats at the tables throughout the room, ready to be paired in one-on-one scripted conversations.

But first the mixed group did a quick get-to-know-you exercise by telling one another about a favorite meal. When it came out that a service warrior named Darrell recently tried broccoli for the first time, the well-meaning corporate volunteers erupted in applause. It was a bonding moment, something to break the ice. I found this charming in a benign way, but my eyes and ears were also on Jeff. He knew that the applause was also the subtlest evidence of a stubborn idea in the minds of these employee partners: that these near-adults are forever-children. Their impulse was to celebrate a young man venturing out in his tastes and habits as though he were half his age. Jeff moved the group along to the interview activity at hand, but not before he delivered a casual aside, loud enough for everyone to hear. He was smiling when he said it, but the sting was unmistakable: "I can't believe you all clapped for that."

W e already accept that most couples don't want a Down child," said James Watson, a geneticist who was deeply involved with the discovery of DNA and the Human Genome Project. "You would have to be crazy to say you wanted one, because that child has no future." No future, meaning that child will be a forever-child, with no temporal horizon, no imaginable life that grows and changes and matures in the long arc of time. That's the assumption, anyway—that having some basically permanent needs for support automatically renders a person in a permanently stunted state. "Once a child and always a child" is what people say; it's what people have unthinkingly, casually said to me. Watson is known for mouthing off about social topics just for sport, in ways most people would find alarmingly objectionable. But here he's speaking something many people think but won't voice aloud—that the prospect of having a child who lacks normative intelligence fails the most important demand of the clock. The idea of a person without established "high functioning" economic worth just doesn't seem like someone who will access a meaningful future.

I can only think that this blank, futureless prospect is partly driving the global majority of parents to opt out of having children with Down syndrome via selective termination—the most invisible and everyday form of eugenics, shaping and culling the future by means of one privately held, economically distinct family decision at a time. In my own country, where healthcare isn't guaranteed and social infrastructure is unevenly distributed, many families may well terminate because they lack supports for raising a child with disabilities. I've talked to and read about many women who conceived a fetus with Down syndrome and decided to terminate; there are dozens of reasons why they do so. Some feminist philosophers frame these decisions—whether to birth and raise any child or terminate a pregnancy—as life choices that defy the category of mere "choice" itself. Instead, they say, these decisions are each options that play out in a wider frame of "reproductive justice" that includes matters of rights but involves much more. Choices about pregnancy are inherently contextual, involving deep racial and class-based differences and inequities. A vision of reproductive justice, first articulated by alliances built by women of color, includes reproductive choices not only to conceive and birth children, but also to raise them and educate them for the long term in cultures where they may thrive.

Still—perhaps paradoxically—in much of northern Europe, where social safety nets are strong, the termination rate for fetuses with Down syndrome is actually a higher majority (in some cases, upward of 80 percent) than in the United States (around 68 percent). There are many places where shared cultural stories about desirable futures just don't imagine Down syndrome in them, where normative intelligence has become a necessary feature in the story of overall human health, and non-normative intelligence, in turn, has been conflated with disease. Whatever the reasons, selective termination for Down syndrome is a collective design process happening one decision at a time, a process that is homogenizing human populations by genetic selection. That's why a geneticist like Watson was commenting on Down syndrome in the first place: because the genetics of Down syndrome are easy to spot and unusually clear. It's an on-or-off switch; either you have it or you don't. And that information is detectable by another kind of assistive technology, the testing regimen that is part of the reproductive health process, the very ordinary steps in a woman's normative prenatal visits to the doctor.

So how does a pro-choice mother like me voice a simultaneous claim against the silent and even unwittingly eugenic process happening in contemporary pregnancy? How does a mother support the worth of her child living on an alternate clock? There are no easy answers. Many days I feel stymied into silence by the conflict in my own convictions and by the absence of reproductive justice that relegates the matter of selective termination to a narrow definition of private choice. But on those days with the EPIC service warriors, I watched the young people considered forever-children, the ones just a few years ahead of Graham, looking to build their future, young people both aspiring to and at odds with a prescribed notion of adulthood and the economic productivity that is its only goal. I glimpsed, at least, the face of a different clock unfolding to a time somewhere out there, somewhere ahead of now.

A year after the mock job interviews, on a November Saturday, I went to the eighth annual City Serve event that EPIC runs as their biggest production of the year: a Saturday devoted to a big group service project. It's the same job every year—repainting all the doors of a high school somewhere in the city of Boston—and this day's task was to cover all 139 doors at Brighton High. The school was built in 1931, a glorious old stone structure with soaring arched windows, worthy of a cathedral; they lined the cafeteria, where the cheapest possible tables and benches sat below them. The walls in between were hung with banners celebrating the sports records and faded posters with attitude reminders, the kind that always try a little too hard. We gathered in the cafeteria for registration and an opening ceremony, and later, with cans of bright orange paint, the EPIC service warriors ran this massive undertaking, directing the efforts of a hundred other volunteers.

I talked with a service warrior named Jeremy amid the doughnuts and coffee before the day started. Jeremy is a marathon runner and was made an honorary drill instructor with the sheriff's department in his hometown after an internship through Opportunity Works, a job development program for young adults with disabilities. He has the barrel chest and shoulders of the officers with whom he worked over the summer; he also spoke with a stammer and often placed the tips of his fingers behind his ears while talking, eyes darting about but open to chatting. He's in his early twenties, autistic, and now a service leader, so his job was a big one. I'd seen him earlier in the hall, practicing the script of how he would guide the small group who would join him for the day. It was his task to dole out jobs and supplies and generally herd his crew. He told me about the marathons and the sheriff's department and what his tasks for the day would be. "My job is to put people into groups, get them pumped for the day," he said. "EPIC is helping me with social skills, even though I'm popular out there."

I also spent some time talking with Terence, another service warrior, who was nervous about the day's responsibilities. A wheelchair user who flapped and hummed between sentences while we spoke, he sat at his first post at the main entrance and guided people inside to the registration forms—pointing and saying, *Right through those doors and to the left*. A first-time service warrior, he was trying to decide if he'd continue; soon he would be transferring to a residential special education program to finish high school, and he was reconsidering whether to keep giving up so many Saturdays every year. "Weekends are the only time I get to see my family," he said. Two Saturdays a month was a big commitment.

Later in the morning, we were brought to order by the EPIC leadership, including a new service warrior, Emily, a student at Perkins School for the Blind who admitted to nervousness about public speaking. She made her way to the microphone with sunglasses and cane and talked about her appreciation for EPIC as an experience. She joked casually that most of us would "see" the big results of the painting work before the day was done. "But I won't," she said, and chuckled. The cafeteria was one of those cavernous, echo-y spaces, so her voice didn't carry enough for the joke to land. "Have fun," she said, and sent us out to the hallways.

The night before the event, EPIC staff and volunteers had taped and prepped the doors and laid out equipment and squares of plastic below each threshold to catch the drips, so the day itself unfolded easily: sanding, cleaning, and then the painting. It was a big operation that they had down to a science. I walked the halls and observed painting teams, pausing to chat briefly but moving through all three floors to see as much action as I could. Groups of three and four gathered around each door frame. Some service warriors needed spotting to reach the higher areas, if they used walkers; others needed help gripping the sandpaper with enough force.

As I walked, I thought about how fitting it was for this work to be carried out in a public school like Brighton High. The whole building was a mix of that beautiful stone shell and the need for material updates, a living history of my country's formative belief but paltry investment in public education. Many classrooms had the original gleaming dark hardwood floors; they were protected by thick coats of varnish but remarkably handsome—the real thing. The windows were enormous, a dozen feet high, and they were covered by roller shades, bent and sad but doing their reliable job. Occasionally I saw some handmade curtains tied sweetly with a ribbon, the perennial work of teachers everywhere. All the classrooms had charming and very worn dark wood inset shelving; they had similar built-in chests of drawers that flanked each side of the old-school blackboards, storage where books and chalk used to lie. Above each chalkboard perched a 1980s-era Magnavox television, and outside the doors, along the ceilings that lined the hallways, was the completion of this index of history: the extended snaking mass of wires that brought digital connectivity to the campus. This was clearly not a rich school, and the inherited furnishings were there because they had to be; suburban schools would have done away with them long ago. But the mix of the historical and the contemporary bore out visually all the under-construction, long-held ambition a public school must carry—affirming, year after year, the possibility for learning that's due to every single teenager who comes through its doors.

The hallways, too, spoke this mix of past and present. A sober wood memorial hung on the second-floor hallway in homage to the hundreds of men who formed the "Honor Roll" of World War II. Their names were in brass; a star designated those who had lost their lives. The memorial was suspended amid rows of yellow lockers, decades old, badly scratched and pockmarked. Roman-style friezes dotted the walls above the lockers, relics from the original campus for this building; the formal nude figures were accompanied by colorful posters all over the walls, brightly reminding students to practice safe sex. Quotes above the classroom doors mixed the words of Abraham Lincoln and Sonia Sotomayor with those of LeBron James and Steve Jobs, and inside the rooms were rows of shiny desktop computers, or flasks and beakers, or skill-building washing machines and model kitchens. Brighton High is a high-poverty school, so one whole classroom had been converted into a "store" for students to shop for free: clothing and shoes and toiletries and canned goods to take home as needed. My friends from other rich countries in the world would be aghast, but this is public schooling in America, where the mission extends to forms of support far outside academics.

Political theorists and historians will tell you that *democracy* is a verb. It is built, not possessed. It's impossible to simply "have" or "not have" it. Democracy is a practice—iterated and reformed, debated and expanded, recovered and reinvented. And education, moreover, is a "causal force" behind democracy, writes political philosopher Danielle Allen. The ideal result of an education isn't the industrial clock of train times and synchronization. Allen writes that it's not merely market readiness but "participatory readiness" that people need at the end of their schooling, a combination of skills that constitute healthy, robust "civic agency"— "the activity of co-creating a way of life, of world-building." We learn so that we might deliberate, all of us together, about desirable futures. Who belongs in them, and why, and how, depends in part on the stories we tell about each other.

The contexts for world-building can be quite ambitious, Allen writes, at the scale of cities or nations. But they can also be quite modest: in a local neighborhood or in a single school like Brighton High. You might even say that a public school is an *action setting* for democracy. It seemed to be so that Saturday—a collective of a hundred volunteers, led by EPIC service warriors, at work in the unfinished structures of the public school. All these misfit bodies doing the labor of shoring it up with paint, meeting the world and making it new, with and without assistance. I passed by Terence, the one who'd been greeting folks at the door, as he was wheeling the halls. He exchanged glances with a friend, trying to assess his role: "I'm keeping my cool, right? I think I'm doing okay."

MAKING ASSISTANCE VISIBLE.

At their best, public things gather people together, materially
and symbolically, and in relation to them diverse peoples
may come to see and experience themselves—even if just
momentarily—as a common in relation to a commons,
a *collected* if not a collective. —BONNIE HONIG,
PUBLIC THINGS: DEMOCRACY IN DISREPAIR

A graffiti overlay forces a reimagination of the International Symbol of Access—and of disability and accessibility by implication—at an early stage of the Accessible Icon Project.

M any things in our lives are created by designers with the idea that they should be useful without people having to notice them—kitchen tools or software applications, for example, that intuitively suggest the steps for their use without taking too much of our attention or brain power. Their invisibility is meant to be their virtue. But some design is made to be emphatically *visible*, bringing new attention to its subject. That aim was the heart of what became the Accessible Icon Project, the only technically illegal street art I've ever done.

In early 2010, I was heavily pregnant with my third child and four years into a steep learning curve about disability rights, law, history, and more. My husband and I were becoming conversant and connected within a disability community that reached far beyond parents of kids with Down syndrome. We were ushered into the bigger landscape of stories and symbols about disability that are carried in all kinds of places—in TV shows and advertising, in the look and feel of special education classrooms, and in the ordinary features of streets and sidewalks. The street features were particularly alive to me in those years, because I was often pushing two toddlers in a stroller and, later, also carrying a baby in a sling, up and down the curb cuts and into and out of subway elevators. Their bodies, traveling with mine, made me marvel with fresh eyes at something that I'd tacitly known but had never truly seen: evidence of the edited city. Curb cuts and ramps and elevators, infrastructure hard won in the generations preceding mine, had made my own passage through public space smoother. And now I was the parent of a child who would be asking for more and different accommodations, a more flexible world for his growing up. So for the first time, too, I started to really see the International Symbol of Access.

The International Symbol of Access (ISA) is one of the most mundane and powerful symbols in public space: the image of a person using a wheelchair depicted on signs outside a café or city hall or preschool, protecting parking spots or indicating the presence of ramps and accessible doors. The ISA is a standardized and rote image in public, doing its pragmatic work by being readily available and easy to spot, always in the same blue and white tones, internationally legible because it doesn't depend on text of any kind. Its recognizability makes it a symbol for physical access for wheelchairs but also for generalized assistance, physical or otherwise, at the airport or doctor's office. It's what graphic designers call an isotype—a simple two-dimensional graphic that city planners and architects often employ to signal the purposes of built space, like the symbols of human bodies you see on the doors for bathrooms and emergency exits. The standards for what isotypes should look like are formalized and deliberate, because they have to provide maximum contrast, consistent scale in their repetition, and easy legibility even from a distance. The practical details matter.

But the ISA is also an abstract idea made concrete, and that bigger idea gets lost when infrastructure does its handy work of becoming invisible, part of the background. The ISA holds a big promise: an image that permanently, reliably reserves and protects bits of public space for people whose bodies meet the world in a clash. The nearest parking spaces and curbside entrances, guaranteed for those who need them—that's the hardscape, making room. But the symbol stands for something bigger still. It stands not just for these accessible material features of the world but also for everything that material access makes possible: entryways to public life, school and transportation and workplace. The very idea of legally protected access is exceedingly rare in the world and only came to be commonplace in recent decades, and it is still a historical anomaly. A symbol isn't ever intended as a literal picture of a singular person in a singular kind of world. The ISA isn't just for wheelchair users. Instead, it gathers a whole set of ideas. It collects and stands in for accessible goods and services that it captures and holds together, organizing them in a simple, intelligible way.

Still—so much of what becomes ubiquitous also goes to sleep in our consciousness. That's what infrastructure does best, after all: it performs its steady public work ideally without fanfare, only drawing attention to itself when the joints or systems break. The image was a wake-up call to me in those early years parenting a child with Down syndrome. Even though none of us was—or has yet been—a wheelchair user, it stood for physical entry, and it stood for much more. Misfit bodies had been clashing with the built world for a long time, and my son's story was part of this larger narrative that was literally and figuratively under construction. The ISA, together with ramps and curb cuts, was a daily reminder that the features of the built world could be changed, because here was the evidence, right in front of me, that they had been. I started an informal collection of variations on the image by other designers, admiring and comparing their features, and remained frustrated by how invisible each was in the built environment. I wanted to wake these politics once again.

I worked with a graffiti artist slash philosopher, Brian Glenney, to try out some possibilities for a new and altered image of wheelchair access that could sit on top of the old one, precisely to draw attention to it: What might its features look like? How might we emphasize the figure, the symbolic person in the icon, more than the wheelchair? And what would it be made of? Vinyl? Spray paint? We arrived at a sticker with a clear backing, designed at a scale to fit neatly on top of standard signs that featured the original ISA. To fit neatly, that is, but not exactly. We wanted to show the old symbol transposed with a new one, to make it demonstrably, un-ignorably visible once more—to evoke its history and its work as a symbol, to pose the meaning of icons in public as a series of questions in the way that street art has always done. I'd already been influenced by *interrogative design*, a term coined and an idea practiced by an industrial designer and artist at MIT named Krzysztof Wodiczko who later became my mentor. Interrogative design—making things not only for solving problems, but to ask *questions*. That's what street art, stickers or graffiti in the built environment, can do so effectively: you stumble upon its manifestations unannounced, and maybe they catch your eye or even cause you to do a double take. Street art often resists the kind of slick messaging that advertising specializes in. It does its best work when it's got a strong quality of enigma, when it enters the visual field of the street as a surprising bit of grit. That's what we were aiming for. If you were someone who thought the ISA wasn't for you, and you happened upon this new-upon-old layering, would you look at the symbol with new eyes? Would it make you pause to think—not about graphic design, exactly, but about disability, in the street and in the square? Who is the world built for?

We started with a very modest experiment: placing a few dozen (removable) stickers on signs around greater Boston. It wasn't exactly under cover of night—stickers appear on infrastructure all the time—but we were surreptitious about it, vaguely worried about getting caught in the act. It was a test-and-see effort, informal, a design for learning how the idea might be received. We offered some for free to others to test in a few additional locations. We got outsize press for such a small project, and people started asking us for a more formal version of the image—one that would be compliant with the Americans with Disabilities Act regulations and usable on official signage. Our street art campaign morphed into a fully fledged new isotype entirely, requested by and co-designed with many disabled collaborators, expertly rendered by graphic designer Tim Ferguson Sauder, and offered as a symbol for taking up these bigger questions. The new image, now meeting the standards of formal isotypes everywhere, became an icon in the public domain. It's now free for use by anyone; we neither own it nor lobby for its use. The Accessible Icon Project has become a resource employed by others all over the world—on taxicabs in Manhattan and on custom hospital signage in Delhi, on accessible gondolas in Venice and on posters and graphics made by Saudi disability rights advocates. It's used in U.S. government buildings and held in the permanent collections of two museums. It became not just the work of a couple of street artists but a *public* thing, something that has never been a commercial product for us. A lot of people loved it and some people hated it and let us know, as with any public thing. We were surprised by its success and by the criticism. But we were glad, most of all, for the newly awakened debate it raised about disability, a subject of unfinished rights and unscripted futures that must continually be brought into the public eye. Nearly ten years on, the project has mostly exited our authorship. It's a free tool, practical and symbolic, and it belongs to anyone or any organization in the ways they decide—people like and unlike us. But the work began and endures in my mind in that rowdy provisional early stage: as interrogative design.

Like Amanda's lectern, the icon project is a "what if" question in material form, an example of what the late philosopher of aesthetics Maxine Greene called "social imagination." Greene spent her life's work proclaiming the shared power of art in a three-way exchange that happens when two or more people gather and deliberate about some expressive artifact in front of them: a painting or a sculpture or a dance. In that triad, big or small—more than one person, together, in the presence of an object—what's enacted is a glimpse of "conscious possibility," she wrote. Social imagination is "thinking of things as if they could be otherwise."

As if things could be otherwise. Greene was reminding us that imagination doesn't have to be some lavish spectacle or fantasy. It can include those things, but it can also be quite subtle—a sense of the possible in an object or story or performance, brought to our minds and given real attention in the realm of the social, between ourselves and someone else. And surely we can agree—a deliberative, sustained sense of daily *possibility* is vanishingly rare! Every weary realist impulse inside us resists it. But possibilities are "what if" questions, the nourishment we get from being jolted awake by seeing the possible world made new, in tiny gestures or in big ones. By placing the *otherwise* in front of us, posing altered realities in concrete objects, Greene observed, works of art can "evoke an intimation of a better order of things." Sometimes that intimation of *better* happens only by unmasking the hard and ugly truths of the way things are. Sometimes it happens by visualizing a desirable world we want to see. And sometimes that evocation is just in the act of the interrogative. The "what if" question, made by design.

Most of the stories in this book feature practical design and engineering in the lives of disabled people every day, not street art. But Greene tells us that the otherwise is needed here, too: "Our social imagination," she writes, " is the capacity to invent visions of what should be and what might be in our deficient society, on the streets where we live, in our schools." Telling and retelling the stories in this book is another kind of invitation to social imagination: making forms of assistance newly visible, aiding us in seeing the help that's happening everywhere with new eyes. Assistive tools and adaptive designs, at work in our lives and those of others—not all the same, but all connected. These aren't "special" forms of assistance. They're the plainest visible evidence of bodies getting help from the built world and from other people—a state of needfulness that's foundational and even salutary in every life. Legal scholar Martha Fineman writes that the very idea of an individual as a generically self-sufficient citizen, with rights and obligations based on being free from needs and the sole actor of (usually his) own life, is built on an unexamined "autonomy myth" that forms one of the core tacit assumptions of contemporary life, at least in the United States. "We have an historic and highly romanticized affair with the ideals of the private and the individual, as contrasted with the public and the collective as appropriate units of focus in determining social good," she writes. What has been relegated to private life—the realm of the family, with its invisible labors and dependencies of all kinds—might be recovered and acknowledged as a realm for public concern. Assistance, dependence, vulnerability: these embodied experiences have the dignity of the truly human about them. They create networks of caregiving that sustain us all. The designed world may do some work toward making these matters more visible and less exclusionary, but design can be only one small part of a more robust democratic vision of one another—a civic vision of life together, with the body *plus* its many extended needs intact *and* the agency to build, unbuild, and rebuild our worlds, in modest experiments that we test and grow, expand as needed, retool and refine, and perhaps lobby to have supported as social infrastructure.

If assistance is all around us, why does it require so much imagination to see it? The otherwise at work in social imagination might be a call for new tools or environments that reshape the built worlds we live in. But the *otherwise* also might require imagination to grasp the wonder of the help we're already getting. That's the sense of everyday possibility that I've seen in more than ten years of encountering disability at the heart of built things: art and design and engineering. Rather than thinking about disability only as a *problem* to solve, we might engage our wonder, letting "what if" questions grab and hold our attention for a moment, making us rethink what a body can do. Designed things can bring us that bit of productive uncertainty if we let them. They cast new light on the inherited rigidity of normal in a narrow trajectory of industrial time, and they point to the otherwise at hand.

The "what if" questions are what I saw while Chris wrangled a diaper one-handed at the changing table with Felix, and in the hallways of

DeafSpace. They were alive in the cardboard carpentry, down to the headrest for young Niko, and alive between Audre Lorde and her post-op nurse. They were in the tiny cursor on Steve's glasses, in Betsey's new vegetable peeler, Cindy's pen holder, and those lines made by Stephen's tape. Some "what if" questions are useful products—solutions, even, to problems that were earnestly sought out and replicated at scale. Others are singular ideas: an artifact organized around one person in pursuit of otherwise. How might any of these designs be more fortified—by offering their assistance as a matter of *public* as well as private interest?

Making assistance visible calls for design of all kinds—small experiments at the local level and large-scale guarantees when warranted. I watched with interest as new design ideas sprang up with the rising tide of infection under COVID-19. Suddenly, grocery stores redesigned their services to include shopping hours reserved for older adults. My engineering colleagues and other tech professionals debated methods for manufacturing protective face masks with 3D printers or with sewing machines. Still others pursued ways to convert sleep apnea machines or other devices into ventilators. I thought of Roberts and the makeshift dorm rooms created at Berkeley's Cowell Hospital just as universities started assessing their readiness to convert space in the opposite direction if needed: turning their newly emptied dorm rooms into temporary clinics for care. With humility and collaboration, these are ways that designers might work on public things, and not only in times of extremity. In philosopher Bonnie Honig's words, public things can *collect* us for a moment around a possibility or an idea, even if they don't make us into a permanent *collective*. The designs in this book together pose an enduring question—newly augmented by crisis, perhaps, but it was there all along, with disabled people right out in front, asking: Which tools for assistance will we agree to owe each other?

ACKNOWLEDGMENTS

This book and my wider practice have been made possible by decades of scholars and activists in disability studies and disability rights; only a fraction of those texts and thinkers appear in these pages. Michael Bérubé's *Life As We Know It: A Father, a Family, and an Exceptional Child* and Graham Pullin's *Design Meets Disability* arrived in my life when my son Graham was very young. Both texts brought the right words and images at the moments they were needed, helping to launch more than a decade of study, and each of them has powerfully influenced the ideas in this book.

Thanks to Patrick Anderson, Catherine Kudlick, Mara Mills, Georgina Kleege, Susan Schweik, Victoria Marks, Heather Love, and Darrin Martin for their colleagueship, and for an early residency through the University of California Institute for Research in the Arts that connected my work to the active scholarship and artmaking of others. Support from a National Fellowship at New America, a Public Scholar grant from the National Endowment for the Humanities, and an Artist Fellowship from the Massachusetts Cultural Council gave me the miraculous gift of unimpeded paid hours to write. Residencies at Yaddo and the Carey Institute for Global Good provided a rare mix of solitude, community, and sustenance to do the slow-think writing this project required. (Thanks especially to Tom Jennings and the Carey Institute for its nimble redesign to make a family-friendly residency I could manage—a gesture truly in the spirit of the book.) Juhi Bansal and Ahmedabad University generously invited me to teach a workshop that took me to India for research. I loved spending time as a Neighborhood Salon Luminary at the

ACKNOWLEDGMENTS

Isabella Stewart Gardner Museum, participating in a community of Boston-based artists led by the amazing Rhea Vedro; we cheered one another on in our year together. And the Anne McNiff Tatlock Fellowship at Vassar College gave me a test kitchen for the book's ideas with students—led by the visionary Lisa Brawley—and a curious and generative community of scholars who gathered over dinner in a faculty seminar to workshop ideas in disability theory.

My agent, Lydia Wills, saw from far off in the distance what this book could be; her confidence in the project predated mine. Rebecca Saletan proved to be the editor any writer would wish for—patient with the process, always drawing out ideas in their nascent state, and keenly attentive to every line. Thanks to Geoff Kloske and everyone at Riverhead Books, especially Helen Yentus, Jason Booher, Lucia Bernard, Shailyn Tavella, Jo Cunningham, Kasey Feather, Michelle Koufopoulos, Catalina Trigo, and Anna Jardine.

Thanks to my colleagues at Olin College of Engineering—especially Deb Chachra and Lynn Andrea Stein, who suggested some years ago that I might be able to teach engineers, and Vincent Manno, who took a chance and hired me to do so. I deeply value our one big "super department" of interdisciplinary faculty and the professional life we share. This book's dedication is offered with thanks to the students in my class called Investigating Normal, first invented at the Rhode Island School of Design and then run at Olin College. In both institutions, my students took up the topics I offered with enthusiasm and good restless questions. Student assistants Toni Saylor, Mary Martin, William Lu, and Kelly Brennan helped explore the ideas in this book from its early stages, and librarian Maggie Anderson provided astute research assistance and fact-checking.

Sara DeBoer, Lisa Brawley, Deb Chachra, George Estreich, Brian Funck, Jennifer Grant, Elizabeth Guffey, Dotty Hendren, Tim Maly, and Bess Williamson did heroically attentive close reads at crucial formative stages. Matthew Battles, loyal compatriot, agreed to edit repeated drafts and emboldened my voice in the text. My colleague Jonathan M. Adler, ever the Good Doctor, pushed me both to make my ideas more exact and to be more personally present and transparent to the reader.

Thanks to Rosemarie Garland-Thomson, Alison Kafer, David Serlin,

Brian Irwin, Alexandra Lange, Josh Halstead, Matt Correia, Laura Mauldin, Jarrett Fuller, Joel Reynolds, Tamara Morgan, Adam Al-Sawaf, Molly Campbell, and Kate Scully for reading or offering additional feedback and consultation. Thanks to Joanne McNeil, Mimi Onuoha, Jer Thorp, Aimi Hamraie, Kristofer Widholm, Anne Galloway, Kevin Hamilton, Jeff Gentry, Brian Glenney, Alex Zapruder, Diana Berlin, R. Luke Dubois, and my extended Weird Futures crew in the UK. Jeff Goldenson and Adrian LeBlanc gave me the right kinds of permission at the right time. Sandra Zimmerman and Amy Collins have been my two human life preservers for more than a quarter century. Thanks to my FCC community, especially Dan Smith and Kate Layzer, for the practice of faith—for staying with the trouble and singing the darkness—for aspiring to a love that bears and hopes all things.

My mother taught me to love words and languages of all kinds. This book is perhaps above all a work of vernacular translation—between disability studies and design criticism, between the theoretical and the everyday—and I see in it her passion for the complexities of transferred knowledge. And my father, a family doctor who took me along on his hospital rounds in Arkansas when I was very young, taught me never to fear the body in any of its forms. Thanks to them, Dotty and Mike Hendren, and to Alethea and Larry Funck, for all the encouragement.

The playwright Sarah Ruhl wrote a meditation on being a writer and a parent, taking on the much-discussed dilemma of how to find a room of one's own—and the money to support the time—in spite of so much *intrusion* from the likes of children. She found, finally, that "at the end of the day, writing has very little to do with writing, and much to do with life. And life, by definition, is not an intrusion." So I thank my three children—Graham, Freddie, and Malcolm—for bringing so much *life* to my daily, hourly attention, for the radicalizing transformation of becoming a mother and therefore someone who thinks deeply and distinctly about the shapes of parks and sidewalks and bike lanes, about the costs of dependent care, the governance of public school systems, the norms of local street culture, and much more. Most of these goods are held in the commons; they are *public* goods. As I hope this book makes clear, their materials and operations are *designed*, and they have galvanized my love for what political theorists call the "demos"—the public sphere

ACKNOWLEDGMENTS

in which the people might rule. I owe these thanks in the most vividly concrete ways: Thanks to my city's long-held commitment to subsidized, sliding-
scale costs for quality childcare, from preschools and after-schools and
summer camps in the younger years to nearly free teen programs that extend
even to eight o'clock on weeknights for families who need it, and a Mayor's
Office–sponsored summer employment program for those teens to work in
all levels of city infrastructure, seeing and shaping the *action settings* of their
own lives. My community's assumption that parents—not just rich parents—
have all kinds of demanding jobs and working hours is evident in these provisions, and this book is in part a product of that public faith. The fact that
strong, joyful, community-wide investments in *all* young people are a rarity
in my country will forever astonish me.

It was my husband and co-parent, Brian Funck, who first taught me to
turn a skeptical eye toward my natural preference for pristine abstractions,
and instead to pay much closer attention to stories: what came before and what
happens next, one scene at a time. He lives by the beats, the characters, the
surprising turns. It has been my life's singular good fortune to have my own
story, now more than half my life, bound together with his.

NOTES

INTRODUCTION: WHO IS THE BUILT WORLD BUILT FOR?

4 **a piece of furniture:** Amanda calls this project an "Alterpodium" and has written about it formally in Cachia (2016). My description here draws from general insight in Amanda's own narrative. I use *lectern* here to specify the object itself; a podium is typically the larger raised platform in a room for speaking. With Amanda's full analysis in mind, it actually makes metaphorical sense to call her entire project a podium, but for clarity in this book, *lectern* is the more useful noun.

8 **"utility and significance":** Heskett, 26. The chapter "Utility and Significance" lays out an exploration and case studies for how these two concepts are realized in different ways.

9 **an emphasis on the gee-whiz quality:** Disability scholar Ashley Shew calls this phenomenon "techno-ableism." See Shew.

9 **they fly the flag of their help:** See more on specious framings of disability as a problem in Titchkosky and Michalko, 127–29.

10 **Where could statistics give doctors insight:** Cryle and Stephens, 84–85. In the course of the nineteenth century, these questions were a hotly contested matter of debate. I have summarized for the broad arc of time for the general reader.

11 ***normal* referred to being perpendicular or *square*:** Davis (2013), 1–2.

11 **"Once a rule of normality had been applied":** Cryle and Stephens, 215.

11 ***aggregative fallacy*:** Thanks to my colleague Jonathan Adler for assistance in thinking through the use of normal in groups versus individuals.

12 **the average conjoined to the desirable:** Cryle and Stephens, 12–13.

12 **into a paradoxical ideal:** Davis (2013), 1–2.

12 **the "right" kinds of individuals:** See more about the history of eugenics in the "Clock" chapter.

12 **"Some People Are Born to Be a Burden on The Rest":** This history of eugenics and the state fair is drawn from Estreich (2019), chaps. 2 and 3 (sign quotation, 34).

13 **"And the rest is history":** Davis (2002), 39.

13 **scores produced by high-stakes testing:** See also the much more intersectional ramifications linking disability, gender, and race in this long-inherited logic in Erevelles and Minear.

14 **one billion people:** WHO World Report on Disability. The report lays out how its numbers were counted, and also the inevitable complexity of representing these numbers with so many variations in how nations count and assess disability and how sociopolitical factors contribute to disability. For example, the report acknowledges that "many children drop out of school in Brazil because of a lack of reading glasses, widely available in most high-income countries." To mitigate these complexities, the report supplements its two global data sources with the dimensional specificity of regional studies and country-specific surveys.

16 **"a square peg in a round hole":** Garland-Thomson (2011), 593.

16 **her particular tango with dishes and doorbells:** Garland-Thomson dictates her email by voice, too, using high-performance software on her laptop. Her standard email signature is not an apology for the typos that come with dictation in speech-to-text. Instead, she writes: "Thank you for reading flexibly and creatively."

18 **turn his cane into a musical instrument:** See the work of wheelchair dancer Alice Sheppard and Kinetic Light, and the work of Carmen Papalia for more information on these projects, which now have full lives of their own in these artists' practices.

25 **This out-of-sync object:** Eventually, wear and tear took its toll on my class's design, and Amanda sought out a 2.0 version. Designer Hugo Pilate worked with her on a high-performance plastic revision, which is the model Amanda uses now.

26 **the human animal is co-extensive with its tools:** This idea about humans and tools is the basis of cyborg theory, which I touch on in the "Limb" chapter.

27 **"the unexceptional state of existence":** Davis (2007), 4.

27 **"It's too easy to say, 'we're all disabled'":** Davis (2002), 276.

28 **awaiting reconsideration and redesign:** Bruno Latour (2009) has posited that perhaps all design may be thought of as redesign.

28 **"A body's structure":** Deleuze, 218.

29 **"What's more, its composition":** Deleuze, 222.

29 **adaptive, responsive instruments:** In this book I use *body* in ways that scholars would more precisely call *embodiment*—the interaction between our physical corpus and all the possibilities and meanings held in the world external to it. It's a hard word to get right: the common ways many of us talk about bodies is as a thing that we *have*, when really it is a phenomenon that we *are*, and we *are* in ways that change, depending on the shapes of the world.

29 **"What does disability *do*?":** "It is important to think about what disability *does* rather than simply what it *is*." Amanda Cachia, 2012, exhibition essay for *What Can a Body Do?* held at Haverford College, http://www.amandacachia.com/writing/what-can-a-body-do-2/.

31 **the narratives we tell about our lives:** See much more about narrative and disability in Estreich and in Adler et al.

31 *"from the vantage point of the atypical"*: Linton, 5.

32 **what can a body do:** Before Amanda's exhibition by this name, it's originally thanks to Judith Butler, in her conversation with Sunaura Taylor, in the *Examined Life* documentary by Astra Taylor that I discovered disability's connection to Deleuze and his essay "What Can a Body Do?"

LIMB.

33 **"Neither the naked hand nor the understanding":** "And as the instruments of the hand either give motion or guide it, so the instruments of the mind supply either suggestions for the understanding or cautions." Bacon (1620), https://en.wikisource.org/wiki/Novum_Organum/Book_I_(Spedding).

37 **a ritualized social exchange:** It's worth noting that not all prosthetics users welcome a constant barrage of questions about their bodily gear. As my students learned with Amanda, relationships that make questions welcome are built, are earned, and depend on the individuals involved.

39 **questions for enthusiasts of the posthuman:** Katherine Ott's introduction to her co-edited essay collection *Artificial Parts, Practical Lives: Modern Histories of Prosthetics* is a strong call to enthusiasts of the cyborg to keep hold of the physical materiality of both bodies and their gear. Rather than lofty discussions about the posthuman as a mere idea, Ott calls for readers to "keep prostheses attached to people," a productive move that will rightly "limit the kinds of claims and interpretive leaps a writer can make," but "more important, [will] excite new narratives and produce unconventional knowledge." Ott, 2. That book as a whole is a deep inspiration for this one.

39 **"frighteningly inert":** Haraway, 163.

40 **speculation and even fantasy:** This is a tendency in academic research, too. "Most of the scholars who embrace the prosthetic metaphor far too quickly mobilize their fascination with artificial and 'posthuman' extensions of 'the body' in the service of a rhetoric . . . that is always located *elsewhere*—displacing and generalizing the prosthetic before exploring it first on its own quite extraordinarily complex, literal (and logical) ground" (Sobchack, 21).

41 **More than twenty thousand Union amputees:** I owe this history to Matt Coletti's research under Katherine Ott in the archives of the National Museum of American History. https://americanhistory.si.edu/blog/heroes-come-empty-sleeves-0.

41 **uncomfortable to wear:** Still, some 150 new patents for prosthetic arms and legs were filed in the years after the war's end. The legs especially took on new improvements at the end of the nineteenth century: a more responsive bending knee joint, for example, and some spring action in the heel for more natural walking. See Figg and Farrell-Beck for the mixed affordances in this history of invention.

43   **a state of capacity they had once held:** Serlin, 21–56.

43   **"larger civic ends":** Serlin, 12. *"Social engineering of the future"* is emphasized by Serlin.

43   **a future with normalcy restored:** I owe this history, insightful interpretation, and the Stiker reference to Williamson, 19–21.

43   **its economic and patriarchal future:** Aimi Hamraie argues that the entire social model of disability, pointing away from the body and toward the environment, has often been housed in a normalization project toward an outcome of economic productivity— an ironic origin story for what has also brought about a liberatory set of design practices. See Hamraie for more.

44   **a fully "replaceable you":** *Replaceable You* is the title of Serlin's book, a history that weaves together postwar prosthetics with developments in plastic surgery and artificial organs in an era devoted to personal transformation for national ends.

51   **The condom, for instance, was assumed to be outmoded:** Edgerton, 22–25.

51   **Use-centered history . . . "yields a global history":** Edgerton, xi.

54   **infrastructure, local histories, and social norms:** In fact, my entire trip to India had me thinking about parts and systems. I ate some of the most incredible food of my life over nine days in Ahmedabad, and all of it made without meat. The relatively conservative Hindu culture in the whole state of Gujarat is still overwhelmingly vegetarian; you'd be hard-pressed to find meat in restaurants there at all. My American friends love to imagine that eating a plant-based diet is an affectation of rich Western people, but lots of places in the world live with the light planetary footprint of meat-free, affordable, and delicious meals.

55   **"If I am to be [seen as] a cyborg":** Kurzman, 78.

56   **"Ten days after having":** Lorde, 59–61.

60   **"I don't do change":** Personal interview (2016). My colleague in anthropology, Caitrin Lynch, and I collaborated on a digital archive collecting Cindy's many adaptations, viewable at engineerinathome.org.

CHAIR.

72   **It was a lightbulb moment:** These details are drawn from interviews with Truesdell on the ADA's website and from an interview with Truesdell in Lange.

74   **"all sitting is harmful":** Cranz, 18.

74   **harmful enough to shorten life expectancy:** Just a sample of the many studies looking at sitting and health outcomes include: Michelle Kilpatrick et al., "Cross-Sectional Associations between Sitting at Work and Psychological Distress: Reducing Sitting Time May Benefit Mental Health," *Mental Health and Physical Activity* 6, no. 2 (June 1, 2013): 103–9, https://doi.org/10.1016/j.mhpa.2013.06.004; and Ryan David Greene et al., "Transient Perceived Back Pain Induced by Prolonged Sitting in a Backless Office Chair: Are Biomechanical Factors Involved?," *Ergonomics* 62, no. 11 (November 2, 2019): 1415–25, https://doi.org/10.1080/00140139.2019.1661526. Neville Owen et al.,

"Too Much Sitting: The Population Health Science of Sedentary Behavior," *Exercise and Sport Sciences Reviews* 38, no. 3 (July 2010): 105–13, https://doi.org/10.1097/JES.0b013e3181e373a2.

74 **"Biology, physiology":** Cranz, 23–24.

75 **called the "klismos":** See Cranz and also Witold Rybczynski, *Now I Sit Me Down: From Klismos to Plastic Chair: A Natural History* (New York: Farrar, Straus and Giroux, 2016).

75 **an expression of authority:** Opsvik, 22.

75 **"Chair-and-table culture":** Cranz, 29. Cranz also notes that non-Western cultures also use chairs, sometimes for status and rank purposes as well. But in mainland China, for instance, at the time of her writing, chairs were present but not predominant: the backless stool was still more commonly used for everyday sitting than the thronelike, planar-backed model.

76 **"they shape us":** Cranz, 15. Churchill's line that "we shape our buildings; thereafter, they shape us" is from a 1943 debate speech about the rebuilding of the House of Commons chamber after its destruction by bombing. https://api.parliament.uk/historic-hansard/commons/1943/oct/28/house-of-commons-rebuilding.

76 **the Tripp Trapp for children:** See Lange, 74–77.

77 **"Today, industrial design":** Papanek, ix.

77 **"concentrate exclusively on dermatology and cosmetics":** Victor Papanek, *Design for the Real World: Human Ecology and Social Change* (1971; London: Thames & Hudson, 2004), 241. I also owe the analysis and comparison here to Paul Hawthorne's interpretation (https://www.metropolismag.com/ideas/rereading-design-for-the-real-world/).

78 **objects whose very reason for being was to provide safety:** Papanek (1971), 75, 76.

78 **"A new secretarial chair":** Papanek (1971), 92–93.

80 **"we have designed for the majority after all":** Papanek (1971), 56–57.

80 **disabled people have understood "murderous" design:** Thanks to Bess Williamson for this succinct critique of Papanek's inattention to real disability politics: "Papanek [and other socially engaged designers] revealed a design profession eager to explore and appropriate the experience of disabled people for their own learning, but they gave no indication of an awareness of a disability rights movement or other accessible design advocacy. . . . Their emphasis on the benefit of disability awareness 'for us' countered the arguments of the period that access was too expensive to justify, but it also left unexamined the marginal status of disabled people as well." Williamson, 172.

82 **The Aeron was a wild success:** I owe this history of the Aeron chair to Cliff Kuang, "The Aeron Chair Was Originally Designed as the Perfect Seat for Granny," *Slate*, November 5, 2012, https://slate.com/human-interest/2012/11/aeron-chair-history-herman-millers-office-staple-was-originally-designed-for-the-elderly.html.

82 **barrier-free and inclusive design:** The histories of these terms are distinctive from one another. See Hamriae, Guffey (2017), and Williamson for the nuanced treatment of each of these movements.

82 **clues to suboptimal conditions:** For a lengthier introduction to the ideas of universal design in practice, see Holmes.

82 **design "for everyone":** My treatment of universal design is introductory here, intended for the lay reader. Hamraie offers a much more nuanced and complicated history of the various permutations of "universal" design, especially the way it mostly disregarded racial and class distinctions in its claim "for all."

83 **"usable by all people":** Mace cited in Williamson, 147–48.

83 **principles like "simple and intuitive use" . . . "tolerance for error":** Cited in Hamraie, 8.

84 **the idea that he and Betsey hatched:** It's thanks to Liz Jackson's interview with Betsey Farber that she is now re-centered in the origin story of the OXO peeler, which I summarize here. Too often the source of ingenuity is written out of its legacy, as was Betsey in this history, and Jackson's correction of the record is exemplary for publicly restoring the rightful place of disabled people in design. https://www.nytimes.com/2018/05/30/opinion/disability-design-lifehacks.html.

85 **looking closely at disability:** I owe these examples to Williamson, 173–78.

85 **make life better—without your even noticing it:** This description is a quote from designer Davin Stowell at the Smart Design firm, a partner on the OXO Good Grips line with Sam and Betsey Farber, told to me in an interview.

85 **transmittable by electronic means:** Media historian Mara Mills writes that "[Alexander Graham] Bell's experiments with a phonautograph, using a human tympanic membrane to transmit sound waves to a stylus and then to a plate of smoked glass, were part of the foundation for the electrical transmission of speech." See Mills (2010) for more on telephone tech, compressed speech signals, and the long history of computing. It's worth noting that Bell's wish was a charity-minded one, intended to "eradicate" deafness as a disease model of difference. The contribution to technology still stands as a significant one.

85 **closed captioning has become a standard feature of daily life:** Thanks to Larry Goldberg for this history; Goldberg was director of the Media Access Group at the public media station WGBH and a leader in Decoder Circuitry Act advocacy effort.

86 **the key to building a desirable world:** Thanks to Williamson, especially 183–84, for the particularly sharp analysis that helped me think this through.

87 **"as little complexity as possible":** These quotes and narrative are drawn from Winter's own TED Talk, an interview on the TED Radio Hour, and his lab's literature.

87 **think about mass production differently:** ADA also has a connection to the longer history in DIY approaches to disability technology. For the history of DIY disability gadgets and home objects, see Williamson's chapter "Electric Moms and Quad Drivers: Do-It-Yourself Access at Home in Postwar America."

89 **"diffuse design":** Manzini, 13–14.

90 **"answers that change the questions themselves":** Manzini, 13–14.

90 **"cosmopolitan localism":** Manzini, 2.

92 **"One pilot might have":** Rose, 4.

ROOM.

95 **"A house that has been":** Bachelard, 27.

101 **the result of collaborative research:** See Byrd for the description of how the co-design was carried out.

101 **a whisper or a loud demand:** For a beautiful example of how one sign takes on a dozen different meanings, see Christine Sun Kim's TED Talk "The Enchanting Music of Sign Language."

103 **deafness as a way of being:** Malzkuhn cited in Byrd, 245.

103 **the "spatial kinesthetic":** Hansel Bauman cited in Byrd, 242.

103 **The unidirectionality, even "linearity," of spoken:** This "linearity" of spoken English was described by a unnamed Gallaudet student in the commentary of scientist Derek Braun, who was interviewed for an article on the ways signs are created for novel scientific concepts. https://www.nytimes.com/2012/12/04/science/sign-language-researchers-broaden-science-lexicon.html.

104 **"deaf gain" . . . "hearing loss":** This is the central argument that drives the title of Dirksen Bauman and Joseph Murray's edited volume *Deaf Gain* (2014).

105 **visibly cascading from its source to the world:** Thanks to my colleague Mark Somerville for providing background on the mechanics of sound. One of the best things about working in an engineering school and not being an engineer is getting a continuous do-over, a constant beginner's mind, when it comes to my own lifelong education.

106 **"'the present inferiority of the deaf would entirely vanish'":** Cited in Baynton (2015), 49.

106 **oralism as the default mode:** Baynton (2015), 48–51.

106 ***Deaf* as an identity:** Baynton (2015), 48–51.

107 **a vibratory doorbell:** I owe this historical example to Hurley (2016).

110 **newly foreground this piece of who they are:** For a discussion of how adult signers newly acculturate to deaf communities, see, for example, "The Lived Experience of Adults with Hearing Loss as They Acculturate into Deaf Communities - ProQuest." Accessed January 2, 2020. https://search.proquest.com/openview/9f43eec4e5ef49f7b62cad9b8f06db28/1?pq-origsite=gscholar&cbl=18750&diss=y.

113 **"resting on a cloud":** Steven E. Brown, *Ed Roberts: Wheelchair Genius* (Institute on Disability Culture, 2015), 12.

114 **"Helpless Cripple Attends UC Classes":** *Berkeley Daily Gazette*, December 5, 1962.

115 **"We started talking about our rehab experiences":** Willsmore, quoted in Williamson, 100.

115 **"an employer for six years":** Hessler, quoted in Williamson, 100.

115 **newly awake to political strategy:** An attendant recalled later of that period that "there wasn't one square inch . . . that was not seething with the potential of being political." Williamson, 102.

115 **political meetups for "crips and walkies":** *Crip*, then and now, is a political term used as self-description by disabled people. I get into this word more in depth in the "Clock" chapter.

116 **they proposed and won a twenty-five-cent increase:** Williamson, 102.

116 **understood instead as "self-determination":** Kittay (2015). See also Williamson, 97–98.

116 **that person would be less dependent:** This example is paraphrased from DeJong, 24.

117 **"not contingent on a normal body":** Heumann, quoted in Kittay (2011), 50.

118 **Centers for Independent Living now exist:** "Centers for Independent Living | ACL Administration for Community Living." Accessed January 2, 2020. https://acl.gov /programs/aging-and-disability-networks/centers-independent-living.

123 **Call it anticipatory design:** *Anticipatory design* is a term that's been used by some architects and engineers in the past, notably Cedric Price and Buckminster Fuller, but not so far as I know with respect to health or disability; it's usually invoked in conversations about sustainability or desirable (normative) futures.

125 **"It is a fact that humans all have a period":** Kittay (2015), 55.

125 **"It is part of our species typicality to be vulnerable to disability":** Kittay (2011), 56.

126 **"When we recognize":** Kittay (2011), 57. See also physician Atul Gawande's book *Being Mortal*, where he writes that "our reverence for independence takes no account of the reality of what happens in life: sooner or later, independence will become impossible. Serious illness or infirmity will strike." We know this to be true, and yet we're hasty to find easy ways to avoid dwelling on the possibilities. But there's an opportunity in making peace not with just independence, redefined, but also dependence as a fundamental fact. That's the moment, as Gawande writes, that "a new question arises: If independence is what we live for, what do we do when it can no longer be sustained?" Gawande, 23.

127 **"I have received from my daughter":** Kittay (2011), 57.

127 **"People do not spring up from the soil like mushrooms":** Kittay et al. (2005), 443.

127 **"the fundamental aspect of human embodiment":** Snyder et al., 2.

128 **"rather as an alien condition":** Snyder et al., 2.

129 *action settings:* Goldhagen, 183–218.

STREET.

138 **complicity with the Nazi eugenics program:** Baron-Cohen, 305–6.

139 **autism is a spectrum in the fullest sense:** Silberman, 13–14.

142 **Urban planners call these paths "desire lines":** See interview at https://www .nytimes.com/2003/01/05/weekinreview/ideas-trends-whose-sidewalk-is-it-anyway .html.

142 **Urban planners can imagine the lines of use:** Urban planner Riccardo Marini describes his work this way at https://www.theguardian.com/cities/2018/oct/05/desire-paths-the-illicit-trails-that-defy-the-urban-planners.

143 **In Wichita, Kansas, a set of plungers:** https://peopleforbikes.org/blog/plungers-work-after-anonymous-stunt-wichita-makes-bike-lane-protection-permanent/.

144 **not only by efficiency but by *desire*:** There's some urbanism scholarship about this phenomenon. See, for example, Furman, 23.

144 **"The social life of city sidewalks":** Jacobs, 83.

145 **if you can get down the sidewalk:** Susan Schweik covers the fraught history of disability in public in her book *The Ugly Laws: Disability in Public*. She writes that most historians and theorists of urbanism neglect the idea of disability in their analyses of "the right to the city"—the affirmation that city structures belong to everyone. "The body of the city comes to life when it actively and deliberately copes with the *resistance* of impairment and of disability activism," she writes. Urbanists have yet to take up the implications of what this really means (209).

146 **You can see the remnants:** I owe this history to Hamraie, 95–107.

147 **"their reasoning was circular":** Roberts, quoted in Hamraie, 98.

148 **"thick injustice":** Hayward and Swanstrom, 4.

148 **"difficult to assign responsibility":** Hayward and Swanstrom, 4. Thanks to Lisa Brawley for the conversation that led me to Hayward and Swanstrom, and for a dozen other conversations about design and politics that have influenced my thinking in this book.

148 **"When disabled people enact politics":** Hamraie, 95.

148 **Buses were retrofitted:** Hamraie, 95.

149 **Big changes take small experiments:** One of the most famous treatments of this top-down, bottom-up phenomenon of cities is Michel de Certeau's *The Practice of Everyday Life*. De Certeau invites us to think about the work of city planners from the view of a skyscraper—looking down into the street as a system that works for hordes of people and making plans that he calls "strategies." Meanwhile, at the same time, people plan and make the street at the up-close level of the sidewalk down below, what he calls "tactics."

152 **They eventually articulated six pillars:** From executive director Yvonne van Amerongen's 2015 presentation "De Hogeweyk, the Care Concept," https://www.slideshare.net/sherbrookeinnopole/sils-2015-de-hogeweyk-the-care-concept.

154 **often render patients passive:** See Hannah Flamm, "Why Are Nursing Homes Drugging Dementia Patients without Their Consent?," *Washington Post*, August 10, 2018.

157 **a widely varying stance toward truth:** See MacFarquhar.

CLOCK.

163 **growth in the percentage of adults over age sixty:** World Health Organization, *The Global Strategy and Action Plan on Ageing and Health*, 1.

164 **In New York City, crosswalk times:** See the work of New York City's Safe Streets for Seniors program here: https://www1.nyc.gov/html/dot/html/pedestrians/safeseniors.shtml.

165 **five hundred intersections with this technology in Singapore:** Singapore Ministerial Committee on Ageing, *I Feel Young in My Singapore! Action Plan for Successful Ageing,* Singapore Ministry of Health, 2016, https://sustainabledevelopment.un.org/content/documents/1525Action_Plan_for_Successful_Aging.pdf.

165 **"The medical field has a long tradition":** Kafer, 25.

165 **The words *acquired*:** Kafer, 25.

166 **"'only a matter of time'":** Kafer, 26.

167 **"Disability and illness have the power":** Samuels.

168 **"Crip time is flex time":** Kafer, 27.

170 **no etiology of disease:** Down syndrome does come with higher risk factors for physical/medical complications, but not inevitably.

171 **a way to rank humans:** See George Lakoff and Mark Johnson, *Metaphors We Live By* (Chicago: University of Chicago Press, 1980), for analysis about how *high* and *low* are generally regarded as better and worse, respectively.

174 **"Within the walls of the monastery":** Mumford, 13.

174 **"helped to give human enterprise":** Mumford, 14.

174 **"Abstract time became the new medium":** Mumford, 17.

175 **"Railroad time is to be the time of the future":** Cited in Michael Downing's fascinating history, *Spring Forward: The Annual Madness of Daylight Saving Time*, 79–80.

175 **"helped [to] *create the belief*":** Mumford, 15. Emphasis mine.

175 **"more like a kiss than a stone":** Rovelli, 98.

176 **necessitating formal state intervention:** Yanni, 5. Yanni also points to historian Andrew Scull for the larger economic reasons that explain the phenomenon of asylums. See, for example, Scull, *The Most Solitary of Afflictions: Madness and Society in Britain, 1700–1900* (New Haven: Yale University Press, 1993).

177 **"Idiocy" was recast:** Trent, chapter 1, "Idiots in America."

177 **Historian James Trent:** Trent, xix–xxi. My summary of the history is cursory, of course; Trent's excellent book explores each stage of understanding about intellectual disability in the United States throughout its history.

178 **the eugenics era:** Historian Douglas Baynton writes that the fever pitch of the eugenics movement was in part due to shifts in understandings about time—including the changes in everyday time brought about by industry, as in the expansion of train transportation, with its efficient speed and time tables, that rendered the "problem" of intellectual disability not just as a "defect" but as the backwardness of "retardation." See Baynton (2014), 8–11. Thanks to George Estreich's *Fables and Futures: Biotechnology, Disability, and the Stories We Tell Ourselves* for useful framings of this issue and the reference to Baynton. See also Schweik for more on the street as a site of eugenic thinking. Thanks to

Rosemarie Garland-Thomson's writing on eugenics for this reminder of the conceptual category of disability as a eugenic strategy.

181 **"derivative dependence":** Feminist political philosophers have articulated this extended dependence that arrives for all caregivers. Children (among others) may be classed as "dependents," but that "dependent" status also accrues in part to the people who care for them. The entire family structure takes on some measure of the need for broader economic and physical supports. See Fineman and Garland-Thomson (2011) for more discussion and resources.

183 **the best-designed service:** I am alerting the reader to "service design" as a way of thinking about how design transcends just material things, but it's also worth wondering whether and when "service design" should instead be reclaimed as no more and no less than *politics*. Thanks to Lisa Brawley for helping me bring a more discerning eye to the affordances but also the limits of design.

186 **as a cohort that is politically *connected*:** The very creation of a coalitional mind-set has long been established in disability rights circles. "When disability is defined as a social/political category, people with a variety of conditions are identified as *people with disabilities* or *disabled people*, a group bound by common social and political experience. These designations, as reclaimed by the community, are used to identify us as a constituency, to serve our needs for unity and identity, and to function as a basis for political activism." Linton, 12.

190 **But here he's speaking:** I owe this anecdote and insight about Watson to Kafer, 3.

191 **A vision of reproductive justice:** See the SisterSong Women of Color Reproductive Justice Movement's work in this area, starting in the 1990s, and see Ross and Solinger for more reading in this movement. Thanks to Kimberly Williams Brown for pointing me to these resources and connections within disability studies. See also Alison Piepmeier (2014) for an analysis of selective termination in the philosophical tradition of reproductive justice. The death of Piepmeier early in her career was a deep loss for disability studies. See Rachel Adams and George Estreich's edited volume of Piepmeier's work in their forthcoming *Unexpected: Motherhood, Prenatal Testing, and Down Syndrome* from NYU Press.

191 **the termination rate for fetuses with Down syndrome:** See statistics for the United States in Jaime Natoli et al., "Prenatal Diagnosis of Down Syndrome: A Systematic Review of Termination Rates (1995–2011)," *Obstetrics and Gynecology* 32, no. 2 (February 2012), https://obgyn.onlinelibrary.wiley.com/doi/full/10.1002/pd.2910. For Denmark, see Charlotte K. Ekelund et al., "Impact of a New National Screening Policy for Down's Syndrome in Denmark: Population Based Cohort Study," *BMJ* 337 (November 27, 2008): a2547. Screening policies vary widely; see this comparative study across European nations for more information: P. A. Boyd et al., "Survey of Prenatal Screening Policies in Europe for Structural Malformations and Chromosome Anomalies, and Their Impact on Detection and Termination Rates for Neural Tube Defects and Down's Syndrome," *BJOG* 115, no. 6 (May 1, 2008): 689–96, https://www.ncbi.nlm.nih.gov/pmc/articles/PMC2344123/.

191 **shared cultural stories about desirable futures:** See Estreich for a much richer analysis about the ways that narratives drive many ideas about intelligence and health, including ideas about Down syndrome.

195 **not merely market readiness but "participatory readiness":** Allen (2016a). See also her earlier Tanner lectures and formative responses (2016).

196 **they can also be quite modest:** Allen (2016). "World-building" is Allen citing the original language of philosopher Hannah Arendt.

EPILOGUE: MAKING ASSISTANCE VISIBLE

197 **"At their best, public things":** Honig, 16 (emphasis added).

201 **it stood for much more:** See also a similar idea about the way the ISA belongs to disabled people beyond wheelchair users in Guffey (2018). Guffey (2017) also has a vital history of both the design of wheelchairs and of all the design work of the ISA.

201 **What might its features look like?:** As I've written extensively from the project's earliest days, in Q&A on the icon project's website, in essays, and in interviews, our icon design was not the first to pose an alternative figure for the ISA. Several designers have offered formal updates in the last two decades, and others have informally designed their own signage to fit in the aesthetics of a particular building. Our project brought the interrogative work of street art to the design—the attention to the symbol and all it stands for. See also Guffey (2017) for the more complete history.

203 **became an icon in the public domain:** Thanks to my longtime friend and colleague Tim Ferguson Sauder for doing the graphics work on this, and to a whole host of co-designers, disabled and nondisabled, who got us to the newest version.

203 **as interrogative design:** In those years, with a baby in my arms, nursing at my desk, I read and reread scraps of paper with Wodiczko's words from interviews and essays. He became my mentor when I went back to school at Harvard in 2011; it was the spirit of questions that he inspired that helped me set up what would become my lab for design.

203 **"thinking of things as if they could be otherwise":** Greene (2001). Thanks to my longtime mentor, Steve Seidel, who first introduced me to Greene a quarter century ago.

204 **"evoke an intimation of a better order of things":** Greene (2001).

204 **"Our social imagination":** Greene (2001), 5.

205 **"We have an historic and highly romanticized":** Fineman, xiv.

205 **engage our wonder:** Tanya Titchkosky (2011) calls for an entire "politics of wonder" in thinking more imaginatively about disability.

BIBLIOGRAPHY

Adams, Rachel. *Raising Henry: A Memoir of Motherhood, Disability, and Discovery*. New Haven: Yale University Press, 2014.

Adler, Jonathan, et al. "Identity Integration in People with Acquired Disabilities: A Qualitative Study." *Journal of Personality*, December 14, 2019, 1–29.

Allen, Danielle. *Education and Equality*. Chicago: University of Chicago Press, 2016.

_____. "What Is Education For?" *Boston Review*, May 9, 2016a.

Alper, Meryl. *Giving Voice: Mobile Communication, Disability, and Inequality*. Cambridge, MA: MIT Press.

Anderson, Julie, and Heather R. Perry. "Rehabilitation and Restoration: Orthopaedics and Disabled Soldiers in Germany and Britain in the First World War." *Medicine, Conflict and Survival* 30, no. 4 (2014): 227–51.

Anderson, Michael. "Plungers Work: After Anonymous Stunt, Wichita Makes Bike Lane Protection Permanent." PeopleForBikes, March 10, 2017. https://peopleforbikes.org/blog/plungers-work-after-anonymous-stunt-wichita-makes-bike-lane-protection-permanent/.

Bachelard, Gaston. *The Poetics of Space*. New York: Penguin, 1964.

Bacon, Francis. *Novum Organum* (Aphorisms). Translated by Wood, Devey, Spedding, et al. https://en.wikisource.org/wiki/Novum_Organum.

Baron-Cohen, Simon. "The Truth about Hans Asperger's Nazi Collusion." *Nature* 557 (2018): 305–6.

Bauman, H-Dirksen L., and Joseph Murray. *Deaf Gain: Raising the Stakes for Human Diversity*. Minneapolis: University of Minnesota Press, 2014.

Baynton, Douglas. "Deafness." In *Keywords for Disability Studies*, edited by Rachel Adams, Benjamin Reiss, and David Serlin, 48–51. New York: NYU Press, 2015.

_____. "'These Pushful Days': Time and Disability in the Age of Eugenics." *TransScripts* 4 (2014): 1–21.

BIBLIOGRAPHY

Bérubé, Michael. *Life As We Know It: A Father, a Family, and an Exceptional Child*. New York: Vintage Books, 1998.

Bramley, Ellie Violet. "Desire Paths: The Illicit Trails That Defy the Urban Planners." *The Guardian*, October 5, 2018. https://theguardian.com/cities/2018/oct/05/desire-paths-the-illicit-trails-that-defy-the-urban-planners.

Brand, Stewart. *How Buildings Learn: What Happens After They're Built*. New York: Penguin, 1995.

Byrd, Todd. "Deaf Space." In *Disability, Space, Architecture: A Reader*, edited by Jos Boys. New York: Routledge, 2017.

Cachia, Amanda. "The Alterpodium: A Performative Design and Disability Intervention." *Design and Culture: The Journal of the Design Studies Forum* 8, no. 3 (2016): 1–15.

_____. *What Can a Body Do?* Exhibition catalog. Haverford, PA: Cantor Fitzgerald Gallery, Haverford College, 2012.

Cranz, Galen. *The Chair: Rethinking Culture, Body, and Design*. New York: W. W. Norton, 2000.

Cryle, Peter, and Elizabeth Stephens. *Normality: A Critical Genealogy*. Chicago: University of Chicago Press, 2017.

Davis, Lennard. *Bending Over Backwards: Disability, Dismodernism, and Other Difficult Positions*. New York: NYU Press, 2002.

_____. "Dependency and Justice." *Journal of Literary Disability* 1, no. 2 (2007).

_____. "Normality, Power, and Culture." In *The Disability Studies Reader*, edited by Lennard J. Davis, 1–16. New York: Routledge, 2013.

_____, ed. *The Disability Studies Reader*. New York: Routledge, 2013.

de Certeau, Michel. *The Practice of Everyday Life*. Translated by Steven F. Rendall. Berkeley: University of California Press, 1984.

DeJong, Gerben. "Defining and Implementing the Independent Living Concept." In *Independent Living for Physically Disabled People*, edited by Nancy M. Crewe, Irving Kenneth Zola, and associates, 4–27. San Francisco: Jossey-Bass, 1983.

Deleuze, Gilles. "What Can a Body Do?" In *Expressionism in Philosophy: Spinoza*, translated by Martin Joughin. New York: Zone Books, 1992.

Downing, Michael. *Spring Forward: The Annual Madness of Daylight Saving Time*. Berkeley, CA: Counterpoint Press, 2009.

Dryden, Jane. "Freedom, Transcendence, and Disability: Rethinking the Overcoming Story." Paper presented at joint panel of the Society for Existential and Phenomenological Theory and Culture and the Canadian Philosophical Association, University of Calgary, May 31, 2016.

Edgerton, David. *The Shock of the Old: Technology and Global History since 1900.* Oxford, UK: Oxford University Press, 2011.

Erevelles, Nirmala, and Andrea Minear. "Unspeakable Offenses: Untangling Race and Disability in Discourses of Intersectionality." *Journal of Literary and Cultural Disability Studies* 4 (2010): 127–45.

Estreich, George. *Fables and Futures: Biotechnology, Disability, and the Stories We Tell Ourselves.* Cambridge, MA: MIT Press, 2019.

Feddersen, Eckhard, and Insa Lüdtke. *Lost in Space: Architecture for Dementia.* Basel: Birkhäuser Press, 2014.

Figg, Laurann, and Jane Farrell-Beck. "Amputation in the Civil War: Physical and Social Dimensions." *Journal of the History of Medicine and Allied Sciences* 48 (1993): 456–63.

Fineman, Martha. *The Autonomy Myth: A Theory of Dependency.* New York: W. W. Norton, 2004.

Furman, Andrew. "Desire Lines: Determining Pathways through the City." *Transactions on Ecology and the Environment* 155 (2012): 23–33.

Garland-Thomson, Rosemarie. *Extraordinary Bodies: Figuring Disability in American Culture and Literature.* New York: Columbia University Press, 1996.

———. "Misfits: A Materialist Feminist Disability Concept." *Hypatia* 26 (2011): 591–609.

Gawande, Atul. *Being Mortal: Medicine and What Matters in the End.* New York: Picador, 2017.

Goldhagen, Sarah Williams. *Welcome to Your World: How the Built Environment Shapes Our Lives.* New York: HarperCollins, 2017.

Greene, Maxine. *The Dialectic of Freedom.* New York: Teachers College Press, 2018.

———. *Releasing the Imagination: Essays on Education, the Arts, and Social Change.* San Francisco: Jossey-Bass, 2000.

———. "Thinking of Things as if They Could be Otherwise: The Arts and Intimations of a Better Social Order." *Variations on a Blue Guitar: The Lincoln Center Institute Lectures on Aesthetic Education.* New York: Teachers College Press, 2001, 116-121.

Guffey, Elizabeth. *Designing Disability: Symbols, Space and Society.* New York: Bloomsbury, 2017.

———. "A Symbol for 'Nobody' That's Really for Everybody." *New York Times,* August 25, 2018.

Hamraie, Aimi. *Building Access: Universal Design and the Politics of Disability.* Minneapolis: University of Minnesota Press, 2017.

Haraway, Donna. "A Cyborg Manifesto." In *Readings in the Philosophy of Technology,* edited by David M. Kaplan. Lanham, MD: Rowman and Littlefield, 2009.

BIBLIOGRAPHY

Hawthorne, Paul. "Rereading Victor Papanek's 'Design for the Real World.'" *Metropolis*, November 1, 2012. https://www.metropolismag.com/ideas/rereading-design-for-the-real-world/.

Hayward, Clarissa Rile, and Todd Swanstrom. *Justice and the American Metropolis.* Minneapolis: University of Minnesota Press, 2011.

Heskett, John. *Design: A Very Short Introduction.* Oxford, UK: Oxford University Press, 2005.

Holmes, Kat. *Mismatch: How Inclusion Shapes Design.* Cambridge, MA: MIT Press, 2019.

Honig, Bonnie. *Public Things: Democracy in Disrepair.* New York: Fordham University Press, 2017.

Hurley, Amanda Kolson. "How Gallaudet University's Architects Are Redefining Deaf Space." *Curbed*, March 2, 2016.

Jacobs, Jane. *The Death and Life of Great American Cities.* New York: Vintage Books, 1961.

Jackson, Liz. "We Are the Original Lifehackers." *New York Times*, May 30, 2018.

Jebelli, Joseph. *In Pursuit of Memory: The Fight against Alzheimer's.* New York: Little, Brown, 2017.

Kafer, Alison. *Feminist, Queer, Crip.* Bloomington: Indiana University Press, 2013.

Kittay, Eva Feder. "Dependency." In *Keywords for Disability Studies*, edited by Rachel Adams, Benjamin Reiss, and David Serlin, 54–58. New York: NYU Press, 2015.

———. "The Ethics of Care, Dependence, and Disability." *Ratio Juris* 24, no. 1 (March 2011).

———. "The Personal Is Philosophical Is Political: A Philosopher and Mother of a Cognitively Disabled Person Sends Notes from the Battlefield." *Metaphilosophy* 40, no. 3/4 (July 2009): 606–27.

———, with Bruce Jennings and Angela A. Wasunna. "Dependency, Difference and the Global Ethics of Long-Term Care." *The Journal of Political Philosophy* 13, no. 4 (2005): 443–69.

Kolson-Hurley, Amanda. "Gallaudet's Deaf Spaces." *Washingtonian*, January 2016.

Kuang, Cliff. "The Aeron Chair Was Originally Designed as the Perfect Seat for Granny." *Slate*, November 5, 2012.

Kurzman, Steven. "Presence and Prosthesis: A Response to Nelson and Wright." *Cultural Anthropology* 16, no. 3 (2001): 374–87.

Lange, Alexandra. *The Design of Childhood: How the Material World Shapes Independent Kids.* New York: Bloomsbury, 2018.

Latour, Bruno. "A Cautious Prometheus? A Few Steps toward a Philosophy of Design." In *Proceedings of the 2008 Annual International Conference of the Design*

History Society, edited by Fiona Hackne, Jonathan Glynne, and Viv Minto, 2–10. Falmouth, UK: Universal Publications, 2009.

Lefebvre, Henri. "The Right to the City" (1968). In *Writings on Cities*. Hoboken, NJ: Wiley-Blackwell, 1996.

Linton, Simi. *Claiming Disability: Knowledge and Identity*. New York: NYU Press, 1998.

Lorde, Audre. *The Cancer Journals*. San Francisco: Aunt Lute Books, 2006. Originally published 1980.

Lupton, Ellen, and Andrea Lipps. "Why Sensory Design?" In *The Senses: Design beyond Vision*. New York: Princeton Architectural Press, 2018.

MacFarquhar, Larissa. "The Comforting Fictions of Dementia Care." *The New Yorker*, October 8, 2018.

Manzini, Ezio. *Design, When Everybody Designs: An Introduction to Design for Social Innovation*. Cambridge, MA: MIT Press, 2015.

Meadows, Donella. *Thinking in Systems: A Primer*. White River Junction, VT: Chelsea Green, 2008.

Mills, Mara. "Deaf Jam: From Inscription to Reproduction to Information." *Social Text* 102, no. 28 (March 2010): 35–58.

————. "Technology." In *Keywords for Disability Studies*, edited by Rachel Adams, Benjamin Reiss, and David Serlin, 176–79. New York: NYU Press, 2015.

Mumford, Lewis. *Technics and Civilization*. Chicago: University of Chicago Press, 1934.

Opsvik, Peter. *Rethinking Sitting*. New York: Norton, 2009.

Ott, Katherine. *Artificial Parts, Practical Lives: Modern Histories of Prosthetics*. New York: NYU Press, 2002.

Papanek, Victor. *Design for the Real World: Human Ecology and Social Change*. New York: Van Nostrand Reinhold, 1971.

Piepmeier, Alison. "The Inadequacy of 'Choice': Disability and What's Wrong with Feminist Framings of Reproduction." *Feminist Studies* 39, no. 1 (2013).

Pullin, Graham. *Design Meets Disability*. Cambridge, MA: MIT Press, 2010.

Rose, Todd. *The End of Average: Unlocking Our Potential by Embracing What Makes Us Different*. New York: HarperOne, 2016.

Ross, Loretta, and Rickie Solinger. *Reproductive Justice: An Introduction*. Berkeley: University of California Press, 2017.

Rovelli, Carlo. *The Order of Time*. New York: Riverhead Books, 2018.

Samuels, Ellen. "Six Ways of Looking at Crip Time." *Disability Studies Quarterly* 37, no. 3 (2017).

Schweik, Susan. *The Ugly Laws: Disability in Public*. New York: NYU Press, 2009.

Serlin, David. *Replaceable You: Engineering the Body in Postwar America*. Chicago: University of Chicago Press, 2004.

Shakespeare, Tom. *The Disability Reader: Social Science Perspectives*. London: Cassell, 1998.

Shew, Ashley. "Different Ways of Moving through the World." *Logic Magazine* 5 (2018): 207–13.

Silberman, Steve. *Neurotribes: The Legacy of Autism and the Future of Neurodiversity*. New York: Avery, 2016.

Snyder, Sharon, Brenda Jo Brueggemann, and Rosemarie Garland-Thomson. *Disability Studies: Enabling the Humanities*. New York: Modern Language Association, 2002.

Sobchack, Vivian. "A Leg to Stand On: Prosthetics, Metaphor, and Materiality." In *The Prosthetic Impulse: From a Posthuman Present to a Biocultural Future*, edited by Marquard Smith and Joanne Morra. Cambridge, MA: MIT Press, 2006.

Solnit, Rebecca. *Wanderlust: A History of Walking*. New York: Penguin, 2001.

Titchkosky, Tanya. *The Question of Access: Disability, Space, Meaning*. Toronto: University of Toronto Press, 2011.

————, and Rod Michalko. "The Body as a Problem of Individuality: A Phenomenological Disability Studies Approach." In *Disability and Social Theory: New Developments and Directions*, edited by Dan Goodley, Bill Hughes, and Lennard Davis. New York: Palgrave Macmillan, 2012.

Trent, James. *Inventing the Feeble Mind: A History of Intellectual Disability in the United States*. Oxford, UK: Oxford University Press, 1995.

Van Amerongen, Yvonne. "De Hogeweyk, the Care Concept." 2015. https://www.slideshare.net/sherbrookeinnopole/sils-2015-de-hogeweyk-the-care-concept.

Wendell, Susan. *The Rejected Body: Feminist Philosophical Reflections on Disability*. New York: Routledge, 1996.

Williamson, Bess. *Accessible America: A History of Design and Disability*. New York: NYU Press, 2019.

Winter, Amos. "The Cheap All-Terrain Wheelchair." TEDxBoston 2012 (TED video). https://www.ted.com/talks/amos_winter_the_cheap_all_terrain_wheelchair.

World Health Organization. *Global Strategy and Action Plan for Ageing and Health*. Geneva: World Health Organization, 2017.

World Health Organization and World Bank. *World Report on Disability*. Geneva: World Health Organization, 2011.

Yanni, Carla. *The Architecture of Madness: Insane Asylums in the United States*. Minneapolis: University of Minnesota Press, 2007.